Accessibility in the Laboratory

ACS SYMPOSIUM SERIES **1272**

Accessibility in the Laboratory

Ellen Sweet, Editor
Cornell University
Ithaca, New York

Wendy Strobel Gower, Editor
Cornell University
Ithaca, New York

Carl E. Heltzel, Editor
ACS Committee on Chemists with Disabilities
Consultant, Seneca, North Carolina

Sponsored by the
ACS Division of Chemical Health and Safety

American Chemical Society, Washington, DC

Distributed in print by Oxford University Press

Library of Congress Cataloging-in-Publication Data

Names: Sweet, Ellen, editor. | Strobel Gower, Wendy, editor. | Heltzel, Carl E., editor. | American Chemical Society. Committee on Chemists with Disabilities.
Title: Accessibility in the laboratory / Ellen Sweet, editor (Cornell University, Ithaca, New York), Wendy Strobel Gower, editor (Cornell University, Ithaca, New York), Carl E. Heltzel, editor (ACS Committee on Chemists with Disabilities, consultant, Seneca, North Carolina) ; sponsored by the ACS Division of Chemical Health and Safety.
Description: Washington, DC : American Chemical Society, [2018] | Series: ACS symposium series ; 1272 | Includes bibliographical references and index.
Identifiers: LCCN 2018003574 (print) | LCCN 2018006474 (ebook) | ISBN 9780841232754 (ebook) | ISBN 9780841232761 (alk. paper)
Subjects: LCSH: Chemistry--Study and teaching. | Chemical laboratories--Design and construction. | Chemical laboratories--Safety measures. | People with disabilities--Education. | Educational equalization. | Barrier-free design. | Service dogs.
Classification: LCC QD40 (ebook) | LCC QD40 .A23 2018 (print) | DDC 371.9/045--dc23
LC record available at https://lccn.loc.gov/2018003574

Distributed in print by Oxford University Press

Foreword

The ACS Symposium Series was first published in 1974 to provide a mechanism for publishing symposia quickly in book form. The purpose of the series is to publish timely, comprehensive books developed from the ACS sponsored symposia based on current scientific research. Occasionally, books are developed from symposia sponsored by other organizations when the topic is of keen interest to the chemistry audience.

Before agreeing to publish a book, the proposed table of contents is reviewed for appropriate and comprehensive coverage and for interest to the audience. Some papers may be excluded to better focus the book; others may be added to provide comprehensiveness. When appropriate, overview or introductory chapters are added. Drafts of chapters are peer-reviewed prior to final acceptance or rejection, and manuscripts are prepared in camera-ready format.

As a rule, only original research papers and original review papers are included in the volumes. Verbatim reproductions of previous published papers are not accepted.

ACS Books Department

Contents

Indexes

Chapter 1

Introduction

Ellen Sweet*

Department of Environmental Health and Safety, Cornell University, 395 Pine Tree Road, Suite 210, Ithaca, New York 14850, United States
***E-mail: ems325@cornell.edu**

For some people with disabilities, their interest and skills are best applied to laboratory work. Science laboratories are environments where hazardous materials and processes are in use, and assessments are required to mitigate risk and ensure compliance with Occupational Safety and Health Administration (OSHA) and Environmental Protection Agency (EPA) regulations. Accommodating individuals in a laboratory requires balancing adherence to those regulations, as well as the Americans with Disabilities Act (ADA) technical access standards. Individualized assessment and accommodation are needed to ensure that a qualified individual with a disability can work or study effectively in the laboratory while ensuring a safe working environment for all. This book is intended to be a helpful guide for professionals to understand how to provide equal access to people with disabilities in a laboratory environment. It will review the breadth of protections that are provided by the ADA. This book also covers the roles and responsibilities of persons involved in laboratory oversight, including institutional policies and their limitations with respect to providing appropriate support for individualized assessments in the laboratory.

Introduction

There are many examples of people with disabilities who excel in careers where they work in laboratory environments at universities, at government agencies, such as the National Aeronautics and Space Administration (NASA), and at private companies like Lockheed Martin or Merck (*1*). For these scientists to develop meaningful careers, they must first be empowered to study in our universities. The Americans with Disabilities Act (ADA) ensures that they are able to do this. The ADA requires public (Title II) and private (Title III) educational institutions to provide equal access for qualified individuals with disbailities (*2*, *3*).

The ADA law works to remove some of the pre-conceived notions that many of us hold about what people with disabilities can or can't do. The act ensures that those who are qualified have the ability to demonstrate their skills. The ADA recognizes that while the content of the coursework and the requisite skills that must be developed can't change, the way that those with disabilities approach the necessary work can change. This is where reasonable modification and accommodation come into play. Identifying appropriate accommodations requires taking a look at the individual, the course material, and the specific laboratory space to identify the best options in an ***individualized assessment***. This book will provide insight into how post-secondary institutions can work with existing ADA law to accommodate promising science students with disabilities and provide them all the opportunities they are qualified to pursue.

Learning Objectives

Readers of this book will:

- Be able to recognize the requirements of the Americans with Disabilities Act, which provides equal access in a science laboratory setting to qualified people with disabilities.
- Use increased recognition of accommodations to design individualized assessment plans for people with disabilities who will work in a laboratory.
- Locate resources which support inclusion of qualified individuals with disabilities in laboratory settings.

What the Law Requires

In the workplace, certain groups are protected by anti-discrimination laws. For example the Civil Rights Act of 1964 prohibits discrimination based on race, sex, color, or ethnic origin. The Age Discrimination and Employment Act (ADEA) prohibits discrimination by employers of individuals over 40 years of age (*4*). The ADA is also an anti-discrimination law. It requires that post-secondary institutions provide people with disabilities an equal opportunity to enroll in and participate in the programs and services offered (*3*). Schools are required to provide effective communication, reasonable modification to programs and services they offer, and ensure that individuals with disabilities can access the aids they may need. The law is described in detail in Chapter 2. The architectural standards for these institutions are outlined in the ADA Accessibility Guidelines (*5*). These guidelines provide the minimum level of physical accessibility to the existing work environment. Chapter 3 describes these standards specific for the laboratory setting.

Implementing Accommodation

Chapters 4 through 6 address specific types of disabilities which students enrolled in laboratory courses may have. Each of these chapters provides an overview of the functional limitations of these students and the accommodations that can be used to effectively and safely allow participation in laboratory experiments. Chapter 7 specifically addresses how educators might accommodate service animals in the laboratory environment. The idea of individualized assessments to determine an appropriate accommodation is evident throughout each of the chapters. Accommodation is required by the ADA law and institutional support such as an Office for Students with Disabilities (OSD) should already exist.

In chemical safety, risk assessments, performed by laboratory staff, should be common practice for all new or changing experiments being planned within the laboratory. Faculty and laboratory staff typically have limited (if any) expertise in accommodations for persons with disabilities, even temporary ones. Nor do they have experience with conducting individualized assessment. Staff in OSD programs, who are knowledgeable about the potential implications of specific disabilities and the available options for accommodation, can help conduct individualized assessments which can be included in safety risk assessments. Having both OSD and lab staff collaborate with the student offers an effective way to ensure that students with disabilities are accommodated while providing a safe environment for all students. This also ensures that students with disabilities can maximize their educational results. It is "a logical extension of the culture of safety to include a culture of accessibility" (*6*).

Accommodation in Education while Being Educated Safely

Every learning institution has a responsibility to ensure the safety of their students in science laboratories. These requirements do not change when including people with disabilities.

Specialized knowledge, typically not widely understood by educators, is required in both meeting the needs of those with a disability and minimizing risks from laboratory hazards. "The best safety measures focus on the prevention of accidents" (*6*). The implementation of the measures needed to identify, assess and mitigate hazard for ***all students*** depends on the institutional structure and available expertise. It is critical to ensure that students with disabilities are not presumed to be unsafe and are empowered using institutional resources to safely meet their potential in laboratories.

Risk Assessment in the Lab – Why Is It Needed?

One look at an organization's Health and Safety website might easily overwhelm the OSD professional who likely does not have a science background. For example, an accommodation in the laboratory for a student with a mobility disability must go further than providing an adjustable or lower height fume hood with wheelchair knee space below it. The risk assessment portion of an individualized assessment for a student in a wheelchair must also consider the safety implications of working at eye level with chemicals and protecting the student from spillage because they are in a sitting position.

Another example is the scale up of an experiment. This means the use of larger volumes of chemicals and glassware. Many lab experiments are done at microscale for a variety of reasons, such as to use less chemicals and save on repurchase costs. The risk of exposure to smaller volumes of chemicals may be very different in microscale equipment than if the experiment must be done at a larger scale in order for a person with impaired vision or inability to grasp the smaller glassware who cannot conduct the procedure at microscale. For example, greater volumes of solvents will create a greater risk of fire and more health concerns from inhalation. The fume hood may not be necessary for microscale work, but may be necessary to mitigate the increased risk to chemical exposure at the larger volume.

Risk assessment may involve someone who needs to have their service animal in close proximity in the laboratory. Figure 1 shows a student in the laboratory while her dog looks on. Her dog has been positioned between her as she stands at the fume hood and the weigh station that she will have to walk over to. The risk assessment had to take into consideration the safety of the animal in addition to the needs of the student. Her dog is underneath the counter where he is out of foot traffic and in a location where chemicals are not likely to be spilled on him. The animal has on rubber bottomed booties to protect his pads from chemicals and broken glass on the floor. The risk to the dog has been mitigated, and the student has the full benefit of the service animal while she moves about the laboratory.

Figure 1. A student with her service dog in a laboratory.

The following are a few examples of the safety aspects that the person conducting the individualized assessment for a disabled individual will have to consider. Some others might be:

- When there is a possibility of fire, the emergency response plan must be revised for a vision impaired student. This student may need to practice the route to the closest escape in a non-emergency situation.
- Time of travel for this student, may be longer to the eyewash/shower unit, so their path should be considered as well as their proximity to the unit.

- A student with a disability may have to conduct the experiment in a different manner than others in the lab to be safe. A TA or assistant may be needed.

Institutional Support Structure

Policies and procedures of an institution drive both the culture of accessibility and safety. Due to the myriad of hazards, it is more complicated to be inclusive within the laboratory setting than it is in any other space. The process of creating an accommodation in a lab requires specialized expertise. "Recourses are limited and administration must provide support for teachers who are not subject matter experts" (*7*). In order to truly become accessible, institutions must acknowledge this.

Prudent Practices in the Laboratory states that "laboratory personnel realize that the welfare and safety of each individual depends on clearly defined attitudes of teamwork and personal responsibility and that laboratory safety is not simply a matter of materials and equipment but also of processes and behaviors" (*7*).

The American Chemical Society, Committee on Chemical Safety recognizes the RAMP framework for hazard and risk management in the laboratory. "Management requires method, which in turn requires a framework for understanding situations, environments, and materials" (*8*). RAMP is a management system introduced by Hill & Finster in their textbook for undergraduate students (*9*). RAMP is an acronym for:

Recognize- hazards
Assess- the risks of these hazards
Minimize- the risk of these hazards
Prepare- for emergencies

When conducted properly, the safety risk assessment process is similarly for a person with the disability as it would be for an able-bodied person working in the laboratory, but the staff doing an assessment for a disabled person who will work in a science laboratory must have a unique skill set or work with a team of experts. This process can also enhance the safety of all working in the lab. The final two chapters of this book review effective ways to ensure the lab is a safe place to study for all students. Specifically, they address preserving the essential components of the course of study, and provide an example of a highly developed program, and offer alternatives for institutions to consider when developing its assessment program.

Conclusion

The points articulated in this chapter will be expounded on throughout this book. Each subsequent chapter contains information about the ADA as it applies to postsecondary educational institutions. This information has been provided by staff members at the Northeast ADA Center which is currently located on the Cornell University Ithaca campus. These professionals provide training and advice about this law in a variety of ways, including assisting with the writing of this book. Most chapters were written by a staff member of the ADA Center with a partner author from the Committee on Chemists with Disabilities (CWD) of the ACS. Members of the CWD are professional chemists who may be interested in helping chemists with disabilities or may be disabled themselves and have overcome institutional limitations in order to follow their passion to become scientists. This book is the extension of the symposium titled "Americans with Disabilities Act and Accommodations in the Laboratory" that occurred at the 252nd ACS meeting and was sponsored by the ACS Division of Chemical Health and Safety. Members of this division work to collaborate with experts to provide authoritative technical resources and mentorship in chemical health and safety. The experience and recommendations of these groups is invaluable to institutions seeking to support the persons with disabilities.

References

1. Bonetta, L. Focus on careers: Diversity – Opening doors for scientists with disabilities. http://www.sciencemag.org/careers/features/2007/11/focus-careers-diversity-opening-doors-scientists-disabilities (accessed September 26, 2017).
2. ADA Information and Technical Assistance on the Americans with Disabilities Act, State and Local Governments (Title II). https://www.ada.gov/ada_title_II.htm (accessed October 29, 2017).
3. ADA Information and Technical Assistance in The Americans with Disabilities Act, Public Accommodations and Commercial Facilities (Title III). https://www.ada.gov/ada_title_III.htm (accessed October 29, 2017).
4. Department of Labor Flexible Staffing Arrangements, Anti-discrimination Laws. https://www.dol.gov/dol/aboutdol/history/herman/reports/futurework/conference/staffing/9.7_discrimination.htm (accessed November 5, 2017).
5. United States Access Board, Guide to the ADA Standards. https://www.access-board.gov/guidelines-and-standards/buildings-and-sites/about-the-ada-standards/guide-to-the-ada-standards (accessed November 5, 2017).
6. Pagano, T.; Ross, A. *Teaching Chemistry to Students with Disabilities: A Manual For High Schools, Colleges, and Graduate Programs*, 4.1 ed. (with members of the American Chemical Society Committee on Chemists with Disabilities); Washington, DC, 2015. http://scholarworks.rit.edu/ritbooks/2/ (accessed November 5, 2017).
7. National Research Council, Committee on Prudent Practices in the Laboratory. *Prudent Practices in the Laboratory: Handling and Management of Chemical Hazards*, Updated Version; National Academies Press: Washington, DC, 2011. https://www.nap.edu/catalog/12654/prudent-practices-in-the-laboratory-handling-and-management-of-chemical (accessed November 5, 2017).

8. American Chemical Society Chemical & Laboratory Safety, RAMP. https://www.acs.org/content/acs/en/chemical-safety/ramp.html (accessed September 27, 2017).
9. Hill, R. H., Jr.; Finster, D. C. *Laboratory Safety for Chemistry Students*; Wiley: 2010.

Chapter 2

Running a Laboratory in a Post-Secondary Environment: Regulations Related to Disability

Joseph Zesski and Wendy Strobel Gower*

K. Lisa Yang and Hock E. Tan Institute on Employment and Disability, School of Industrial and Labor Relations, Northeast ADA Center, Cornell University, 201 Dolgen Hall, Ithaca, New York 14853-3901, United States
***E-mail: ws283@cornell.edu**

Individuals with disabilities have often been excluded from or denied full participation in society, education, and employment. Over the last half century, advocacy by these individuals has led to the creation of laws that aim to remove institutional, physical, and programmatic barriers for people with disabilities. Legislation, such as the Rehabilitation Act of 1973 (Section 504) and the Americans with Disabilities Act (ADA) promote the full inclusion and participation of those with disabilities in all aspects of community living including the post-secondary education environment. This chapter seeks to explain how these two laws apply to the college or university setting. It will detail the responsibilities for the institution and for the student with a disability under these laws. It will also provide a framework for understanding Section 504 and the ADA as applied to the classroom and laboratory in colleges and universities.

Why Is Inclusion of People with Disabilities Regulated?

Historically, individuals with disabilities have faced discrimination, isolation, and exclusion from society. While they are not the first group to experience segregation, people with disabilities were still struggling for the legal right to full inclusion long after the civil rights movement of the 1960's. Until recently, there was no protection for people with disabilities who experienced discrimination in their communities, at work, or in involvement in state and local government. As an example, consider the experiences of people with disabilities in employment. People with disabilities are more likely to start in entry level positions, to stay in entry level positions, are paid less, and have fewer opportunities for career advancement (*1*, *2*). In post-secondary institutions, students with disabilities experience physical, social, and emotional barriers when accessing their education including difficulty securing financial resources to pay for college, challenges in receiving necessary accommodations, and lack of physical access to the campus community (*3*). Clearly, the need for legislation to protect the interests of people with disabilities is necessary to ensure non-discrimination.

Disability Rights Protections in Post-Secondary Education

As the disability rights movement geared up in the late 60's and early 1970's, people with disabilities asserted the idea that the problems that led to inclusion was not a result of the limitations of people with disabilities, but rather it was due to the barriers in the environment and the attitudes of others. Where before each individual disability group negotiated their own inclusion, people with disabilities began to band together across disability types to ensure the rights of all people with disabilities (*4*). The first step in establishing protections of the civil rights for people with disabilities occurred with the passage of the Rehabilitation Act of 1973. For the first time, exclusion of people with disabilities was recognized as discrimination. With the passage of this act people with disabilities were recognized as a minority group, or a class, whose rights needed to be protected to ensure equal treatment.

While the Rehabilitation Act has a long history and several sections that protect people with disabilities in various settings, it is Section 504 that impacts education for people with disabilities (*5*). This law was modeled after previous civil rights laws which banned discrimination on the basis of race, ethnicity and sex. Section 504 was designed to protect individuals with disabilities from being discriminated against based solely on their disability in services and programs that receive federal funds such as government agencies, federally-funded projects, K-12 schools, and post-secondary entities (state colleges, universities, and vocational training schools). Litigation and regulations that arose out of Section 504 of the Rehabilitation Act generated central disability rights concepts like "reasonable modification," "reasonable accommodation," and "undue burden." These critical terms are defined in this chapter.

Section 504 of the Rehabilitation Act served as a model for the landmark civil rights legislation for people with disabilities, the Americans with Disabilities Act (ADA) in 1990. The Act represented a combined bipartisan effort on behalf of people with disabilities, the business community, various advocate groups, and the state/federal government. The ADA expanded Section 504 of the Rehabilitation Act to protect people with disabilities from discrimination in a variety of settings, not just those that are federally funded, spanning across the public and private sector, employment, and telecommunications. Among other protections, Titles II and III of the ADA ensure access to post-secondary education for students with disabilities.

As described in this chapter, a post-secondary entity will be covered by both Section 504 and either Title II or III of the ADA. In those instances, compliance with the requirements of both laws must be met. The ADA provides that where two non-discrimination laws apply, the law which affords the greater protection for an individual with a disability applies. Hence a post-secondary institution that is a private entity likely must comply with Title III of the ADA as well as Section 504. In general terms, most of the requirements of Section 504 and of the ADA are relatively similar in nature. Yet when they are different, the student with a disability should be considered in light of the law that offers the greatest protection. We will now turn to consider the specific obligations of Section 504 and of Title II and III of the ADA in relation to colleges and universities.

Section 504 of the Rehabilitation Act

Application of Section 504 of the Rehabilitation Act of 1973 in Post-Secondary Education

The Rehabilitation Act of 1973 states that "no otherwise qualified individual with a disability in the United States shall, solely by reason of his/her disability, be excluded from the participation in, be denied the benefits of, or be subjected to discrimination under any program or activity receiving federal financial assistance" (*6*, *7*). Non-discrimination basically means that universities cannot set exclusion criteria for its programs that screen out, or tend to screen out, people who have disabilities. For example, enrolling in an education program should not require that a student have the ability to walk unaided or have the ability to see or hear without the use of a reasonable modification.

Most (but not all) post-secondary schools receive federal funding in the form of federal grants to pay student tuition or to conduct research activities. A covered institution must meet its Section 504 obligations regardless if it is covered by the ADA or not. A public university would typically be covered by Section 504 and by Title II of the ADA. A private post-secondary institution would be covered under Title III of the ADA. Even private colleges and universities with religious affiliations have obligations under Section 504 if they receive federal financial assistance for students.

Section 504 and the Student

To be protected as an individual under Section 504, a student must be a qualified person with a disability who has disclosed their disability related need. The Rehabilitation Act defines a person with a disability as someone:

- With a physical or mental impairment that substantially limits one or more major life activities;
- Has a record of such an impairment; or
- Is regarded as having such an impairment.

This definition contains three prongs. The first, a physical or mental impairment, refers to the fact that an individual has a condition that substantially impacts a major life activity. "The Section 504 regulatory provision (*8*) defines a physical or mental impairment as any physiological disorder or condition, cosmetic disfigurement, or anatomical loss affecting one or more of the following body systems: neurological; musculoskeletal; special sense organs; respiratory, including speech organs; cardiovascular; reproductive; digestive; genito-urinary; hemic and lymphatic; skin; and endocrine; or any mental or psychological disorder, such as mental retardation, organic brain syndrome, emotional or mental illness, and specific learning disabilities (*9*)." There is no comprehensive list of conditions that qualify as impairments because an exhaustive description is virtually impossible. All impairments must be assessed on an individual basis. Major life activities include such functions as caring for one's self, performing manual tasks, walking, seeing, hearing, speaking, breathing, learning, and working. Major life activities also contain major bodily functions (i.e. endocrine system function, or immune system function). Again, there is no comprehensive list of all major life activities and these examples are for illustrative purposes. The second prong of the definition of disability is that a student has a record of such an impairment. This means that a student may not be impacted by her condition at the current time, but has experienced a condition in the past that may reoccur or that may require a modification now. For example, an individual may have had cancer in the past. Even if that cancer is in remission they likely still require regular health care to ensure the cancer does not reoccur. The third prong of the definition refers to disparate treatment given to an individual. He does not have a disability, but is treated in a discriminatory fashion because of the perception of a disability.

Under Section 504, a student must not only meet the definition of disability, they must also be qualified. In the post-secondary environment, a qualified student is someone who meets the academic and technical standards required for admission in an institution and for participation in a program. This means that a student satisfies the standards required by the university or college for all students. The qualified student is able to perform the essential course activities with or without reasonable modifications.

Finally, in addition to meeting the definition of disability and being qualified, a student must disclose their disability and the need for a related change. This is the right and responsibility of the student. There is no compulsory requirement that a student disclose. If a student chooses not to disclose that she has a disability, however, and the disability is not obvious, then there is no obligation on behalf of the college or university to make an effort to accommodate the individual. For example, if a student experiences poor performance due to her dyslexia but she has never disclosed her disability to the university, she could not claim the university discriminated against her by failing to accommodate her condition. When that same student chooses to disclose, then the institution must engage in the process of determining what academic adjustment might be necessary going forward.

Section 504 and the General Public

While the focus of this book is on the relationship of students with disabilities to the chemistry lab setting, it is worth noting that Section 504, as mentioned earlier, prohibits exclusion of qualified individuals with disabilities or denying them an equal opportunity to receive program benefits and services. This is true for the public as well. Publicly available information must be offered in a means to ensure equally effective access for persons with disabilities as for those without disabilities. A college website may need to be made screen reader accessible so that those with print related disabilities can access the information contained on it whether or not they are students. Recruitment must be conducted in a nondiscriminatory way and not intentionally exclude individuals with disabilities. The application process for the university must be made accessible to those with various types of disabilities.

The intent of the law is to provide equal access and equal opportunity to individuals with disabilities. When looking at the application of the law in a given situation, this concept must be kept in focus. Section 504 does not create preferential treatment for students with disabilities. Rather it seeks to eliminate barriers to participation in education related to a disability.

Obligations under Section 504

For post-secondary institutions, Section 504 requires that they make reasonable modifications to academic requirements, known as academic adjustments, to ensure that qualified students with disabilities are not discriminated against. Additionally, institutions must provide auxiliary aids to individuals with impairments to sensory, manual, or speaking skills (*10*). Collectively, these academic adjustments and auxiliary aids and services are commonly referred to as reasonable accommodations even though this term is not specifically related to Section 504.

An appropriate academic adjustment is a modification to an academic requirement to ensure that a qualified student with a disability has an equal education opportunity. These adjustments can take various forms including; reducing a course load, permitting extra years to complete a degree, providing a note taker or student lab assistant, extended time for tests, modifying a policy to allow a service animal, adaptive lab equipment, modification of the manner in which a specific course is conducted, and others. Requirements that are essential to the program of instruction or to any related licensing requirement need not be altered. Fundamental alterations to a program or adjustments that would cause financial burden need not be granted. Note that undue financial burden should be considered in the light of the institution's overall financial standing. To determine what is essential to a course of instruction, several factors must be considered. Essential course activities are determined by tasks that are fundamental and necessary to complete the course; amount of time spent on a specific task or duty; what the instructor believes to be an essential course activity; and course requirements performed by past and current students in the course (*11*). For example, if a student with a disability wished to be excused from a lab for a course required of all chemistry students, then that may represent a fundamental alteration of an essential program as the lab teaches key skills that cannot be learned in the classroom. In addition, colleges and universities do not need to compromise a code of conduct or ethics. A student with a disability can and should be held to the same level of expectation as their peers. Finally, if a student's accommodation would pose a direct threat, for example, then the institution may deny it. This must be approached very carefully and judged on a case-by-case basis. Factual documented evidence must be available to support such a conclusion. And even when this is the case, the student should be engaged to determine if the risk can be eliminated by provision of some form of accommodation.

In addition to academic adjustments, a university must provide auxiliary aids or services to an individual with impaired sensory, manual, or speaking skills. Auxiliary aids and services can include; sign language interpreting, real time captioning, a qualified reader, and adaptive lab equipment for someone with a manual disability, a screen reader for a classroom computer, etc. Provision of auxiliary aids and services does not include devices or services of a personal nature such as a hearing aid, personal attendant, and readers for personal use.

Implementing the Accommodation Process

The initial step in the academic adjustment process begins with the student with a disability. If the student determines to seek an academic adjustment, then he must approach the appropriate employee or department of the university. Typically, this is an Office for Students with Disabilities or similar department on campus. While the student may work with a particular professor at some point to find the appropriate adjustment, requests should be directed first to staff familiar with the student accommodation process. This can help prevent potential issues if a professor is unaware of disability rights or accommodations proceed in a deleterious manner.

When the student contacts the Office for Students with Disabilities, staff do have the right to ask for and/or require appropriate documentary verification of the disabling condition, such as a doctor's letter (i.e. for physical disabilities), a psychological assessment report (i.e. for learning disabilities or mental illness) or a letter from a high school's special education director. Key items for the office to consider when assessing a request include; how does the disability limit major life activities? And what are the functional limitations that result from the disability? Each accommodation process must be assessed on an individual basis. Post-secondary institution staff, including faculty, should also be aware that a student may have been reluctant to come forward with a request. A student may have wanted to try to succeed without accommodations. He or she may have been embarrassed to ask for assistance. Some individuals may not know they are eligible or may not feel entitled to services (i.e. students with psychiatric disabilities). Faculty should realize that while there is a procedure to follow for a student to request an accommodation, the individual's first thought may be to reach out to the professor. Even if a student comes to the professor with a request for a reasonable accommodation, the professor cannot inquire about the specifics of the student's disability. They should instead be redirected to the campus office that supports student accommodations. In order to create an inclusive environment, faculty should include a statement on their syllabus inviting students who have disabilities to discuss academic needs and then to make certain that the student is proceeding according to the institutional policy.

Enforcement of Section 504

For both the student and the university, there are resources to assist them in navigating this process. The Department of Education's Office for Civil Rights (OCR) both enforces Section 504 and also offers technical assistance to the public. OCR can be found at https://www2.ed.gov/about/offices/list/ocr/index.html or reached at 1.800.421.3481.

The ADA National Network (ADANN), federally grant funded centers that provide technical assistance, training, and research, represent a discrete neutral source of information that has no role in enforcing any laws. In addition to answering inquiries about the ADA, the ADANN can offer guidance on Section 504. All ten regional centers can be found at www.adata.org and all share the same contact telephone number, which is 1.800.949.4232.

Section 504 requires that a covered entity have a grievance procedure established to address concerns of individuals with disabilities who believe their rights have been violated. When a student and the institution are at odds about what is an appropriate academic adjustment or auxiliary aid, the student should be advised how to appeal the decision. If the student goes through this process and believes that his rights have been violated, then he may file a complaint with the OCR. Section 504 is enforced by this office. An individual can file a discrimination complaint and OCR may initiate a "compliance review" of a covered entity. OCR pursues complaints in a variety of areas including, not limited to: admissions, recruitment, financial aid, academic programs, student treatment and services, counseling and guidance, classroom assignment, grading, athletics, housing, and employment. OCR then may investigate the claim to determine its validity and take action based on its findings.

Americans with Disabilities Act

Regulations under Title II of the ADA

Title II of the ADA covers all activities of state and local government entities. In the case of post-secondary education, this means publicly funded colleges or universities. As in Section 504 of the Rehabilitation Act, Title II protects *qualified individuals with disabilities* from discrimination on the basis of disability in services, programs, and activities. The definition of disability is aligned under both the ADA and the Rehabilitation Act, as are the provisions around the definition of "qualified". The regulations laid out in Title II extend the non-discrimination provisions detailed under Section 504 to public entities, whether or not they receive federal financial assistance, in the post-secondary environment, these protections extend to both students and the general public who seek to access programs and services. As in Section 504 of the Rehabilitation Act, eligibility criteria cannot be established that screen out, or tend to screen out, people with disabilities. For example, an institution cannot require that students with disabilities be accompanied by an aid to participate in laboratory settings. Also as under Section 504, a student must disclose a disability and establish its presence with appropriate medical documentation in order to take advantage of the law.

Under Title II, qualified individuals with disabilities must have access to both aspects of the physical environment (as discussed in Chapter 3) and the programs and services offered. Under Title II, full physical access is not required. Rather, a public entity's services, programs, or activities, when viewed in their entirety, must be readily accessible to and usable by individuals with disabilities. What this means is that not every building in a public post-secondary institution must be accessible. However, there must be some accessible buildings on campus so that when a student who requires physical access enrolls in a class or an offering of the university – that class or activity can be held in an accessible location. In practical terms however, this could be virtually impossible for a laboratory and so it is strongly recommended that chemistry labs be considered with this in mind. When a physical location is not accessible, it is essential to ensure that the program or service it offers is still available to a student. For example, if a library on campus is not accessible, it would be acceptable for a librarian to retrieve the information required by a student with a disability as an alternative.

The ADA encourages public entities to ensure that, as much as possible, qualified people with disabilities are able to participate in integrated settings. Separate programs are only permitted where necessary to ensure equal opportunity. Even when separate programs are available, the person with a disability is not required to attend those programs. For example, if a professor offers two sections of the same class. In one, a sign language interpreter is offered for a deaf student. If a second deaf student registers for the second session of the course, the University must also provide an appropriate accommodation for the student in that section of the course.

Title II requires that post-secondary institutions reasonably modify their policies, practices or procedures to avoid discriminating against people with disabilities. For example, if a student with an autoimmune condition required a special respirator to participate safely in a biology lab, it would have to be provided. However, if the public entity can demonstrate that the modifications would fundamentally alter the nature of its service, program, or activity, it is not required to make the modification. For example, a university would not have to excuse a student from a lab required for a degree due to a disability.

Title II of the ADA requires that public post-secondary institutions ensure that communication with people with disabilities is as effective as it is for people without disabilities. In order to do this, they must provide auxiliary aids and services, just as they do under Section 504. Appropriate auxiliary aids and services are often dependent on the length and complexity of the communication involved. For example, to ensure access to class lectures, a sign language interpreter or Computer Aided Real Time Translation (CART) may be necessary. However, lab assignments that require independent interpretation could be given out in writing. Also, as described later in the chapter, electronic information (i.e. websites, Learning Management Systems (LMS), documents, and other written information) have to be made accessible to students with disabilities. Title II entities must allow the person with a disability to voice a preference for communication modalities and give primary consideration to those preferences. According to the Department of Justice (DOJ), "primary consideration means that the public entity must honor the choice, unless it can demonstrate that another equally effective means of communication is available, or that use of the means chosen would result in a fundamental alteration in the service, program, or activity or in an undue financial or administrative burden" (*12*). Interpreters for people with disabilities must be qualified which means that they can ensure the information is being conveyed effectively, accurately, and impartially. Many colleges and universities employ qualified interpreters through ODS or a similar office on campus. They also must designate an individual to oversee Title II compliance. They must also develop a grievance procedure.

Regulations under Title III of the ADA

Title III of the ADA covers places of public accommodation. In the case of post-secondary institutions, this means all private colleges and universities. Title III states that a public accommodation may not discriminate against an individual with a disability in the operation of a place of public accommodation. People with disabilities may not be denied full and equal enjoyment of the goods, services, facilities, privileges, advantages, or accommodations offered by the place of public accommodation (*13*). The non-discrimination requirements of Title III are summarized in three broad principles:

1. Equal opportunity to participate;
2. Equal opportunity to benefit; and
3. Receipt of benefits in the most integrated setting appropriate.

Many of the requirements of Title II also apply to entities covered under Title III. These include the definition of disability, the non-discrimination provisions for qualified individuals with disabilities, the mandate to provide services in an integrated setting and the need to eliminate unnecessary eligibility standards or rules that screen out or tend to screen out people with disabilities. Title III entities are also required to make reasonable modifications to their policies, practices, and procedures to avoid unnecessary discrimination. They are also required to provide auxiliary aids and services to qualified individuals with disabilities. Title III also requires effective communication. Unlike Title II however, Title III leaves the final decision concerning what method of communication is effective to the post-secondary institutions. The law states that the communication method chosen must be effective. This would be better achieved when the institution communicates with the person with a disability to better understand their needs.

As detailed in Chapter 3, Title III of the ADA requires that public accommodations remove architectural barriers (physical features) and communication barriers (conventional signage or alarm systems) that are structural in nature in existing facilities, when it is readily achievable to do so. Readily achievable means easily accomplished and able to be carried out without much difficulty or expense. Under Title III of the ADA, the obligation to provide readily achievable barrier removal is ongoing. That means that something that was not determined to be readily achievable at one point, may be readily achievable in the future. The DOJ has established priorities for barrier removal in Title III entities to facilitate planning for effective access. These include:

- Getting through the door, to include access from public sidewalks, public transportation, or parking;
- Access to places where goods and services are made available to the public;
- Access to restrooms;
- Removing any remaining barriers to access to facilities (i.e. public telephones or drinking fountains).

Just as in Title II, if an entity is not able to remove all physical barriers to access, they must provide alternative access to goods and services. For more information, please refer to Chapter 3 or visit the Title III Technical Assistance Manual available at: http://www.ada.gov/taman3.html.

Additional Considerations

Digital Access to Information

The post-secondary classroom is becoming progressively digital in nature. Consequently, it is important for institutions to be aware of their obligation to provide equal access for all students including those with disabilities. Often, these students may use a form of assistive technology to participate in an online course or other web-based learning. Documents shared by a professor are often in electronic format. Digital accessibility is becoming a larger concern in the post-secondary environment.

Under the ADA, there are no currently explicit established web accessibility standards as of the writing of this book that would apply to a post-secondary educational institution. However, the DOJ has held that web accessibility is intended under the ADA through equal access and effective communication. However, many settlement agreements put into place by the DOJ has aligned to the Web Content Accessibility Guidelines (WCAG) 2.0, level AA, produced by the Web Accessibility Initiative (WAI) of the Worldwide Web Consortium (WC3) (*14*). This standard uses testable statements to determine how well a given website can be accessed or used by individuals with varying abilities based on four principles: perceivable, operable, usable, and robust.

In its role under Section 504, the Department of Education's Office for Civil Rights (DOE-OCR) has not published any official standards for website accessibility. OCR has held that a college or university must provide equal benefits to its programs as well as effective communication including through its websites. However, in several resolution agreements, OCR has suggested Section 508 standards or WCAG 2.0 as accessibility best practice models for universities. These standards now align to the WCAG 2.0, level AA standards. Section 508 is another portion of the Rehabilitation Act that requires all federal agencies to use and purchase accessible information and communication technology.

What might this mean in the college setting? Adopting a general web accessibility policy may be useful in meeting an accessibility goal. Various elements such as electronic documents, online courses, and learning management systems (LMS) need to be addressed. Faculty should be aware that they must create accessible versions of the electronic documents they share with students. For example, Microsoft Word documents are generally accessible to those students that might employ assistive technology. However, images are typically not recognized by screen readers; a form of assistive technology that reads the content of a computer screen or mobile phone. Consequently, images in a Word document must have alternate text describing the image embedded. Faculty can use styles- headings- to improve the navigation features. PDFs (portable digital file) are another common format that can require some intentional planning to make fully accessible. For example, a PDF that is simply a scanned copy of a print document will not work with a screen reader as the program views that PDF as an image. When creating a PDF, it is important to use reading order to help identify the layout. As with a Word document, images within a PDF need alt text. Online courses must be tested for accessibility with assistive technology and for its usability for individuals with various disabilities. LMS similarly must be researched and tested for which platform provides the greatest level of accessibility. A student with a disability needs to be able to access these tools in a manner comparable to their peers without disabilities.

Enforcement activity by the OCR and DOJ bears witness to this expectation. There are numerous settlement agreements, consent decrees, and complaint resolution agreements that touch on the matter. One such instance was a resolution agreement reached by the OCR with Youngstown State University (an ADA Title II and Section 504 entity) in December of 2014 (*15–18*). The OCR found that the school was not in compliance with Section 504 of the Rehabilitation Act and Title II of the ADA. In the first instance, Youngstown State's websites were not readily accessible to persons with disabilities. In the second, OCR found that the university was not fully in compliance with the regulatory requirements regarding the publication of a notice of nondiscrimination in relevant documents. The OCR maintained that they must ensure equal access to educational opportunities for individuals with disabilities including in their web services.

"Accessible" means a person with a disability is afforded the opportunity to acquire the same information, engage in the same interactions, and enjoy the same services as a person without a disability in an equally effective and equally integrated manner, with substantially equivalent ease of use. A person with a disability must be able to obtain the information as fully, equally, and independently as a person without a disability. Although this might not result in identical ease of use compared to that of persons without disabilities, it still must ensure equal opportunity to the educational benefits and opportunities afforded by the technology and equal treatment in the use of such technology.

Among other steps, Youngstown State was required to draft and submit to OCR for review and approval a policy to ensure information provided through the University's website(s), online learning environment, and course management systems "electronic and information technologies" (EIT) were accessible to students, prospective students, employees, guests, and visitors with disabilities, particularly those with visual, hearing, or manual impairments or who otherwise require the use of assistive technology to access information provided through Youngstown's EIT (EIT Accessibility Policy). Moreover, the EIT Accessibility Policy had to identify and adopt a standard by which to measure accessibility compliance. OCR did not specify a single standard, but it gave Section 508 and WCAG 2.0 as sample guidance. In turn, the DOJ has also worked to insure the accessibility of college and university electronic and information technologies. Two such instances include its settlement agreement with Louisiana Tech University and the Board of Supervisors for the University of Louisiana System and its consent decree between Aleeha Dudley and the United States v. Miami University, et al. (*16*, *17*).

Both cases were brought by blind individuals using assistive technology to access LMS and online course materials. In its settlement agreement with Louisiana Tech University, the DOJ held that "all instructional materials and online courses created by a college, department, program, unit or professor must be fully accessible to individuals with disabilities at the same time they are available to any other student enrolled in that program" (*16*).

Both activities by the DOJ required that learning management systems, instructional support applications, and course materials be made accessible. Both specified the use of WCAG 2.0AA as a standard on all new content created and the retro fitting of older materials. Each institution had to appoint a web accessibility coordinator and disseminate a web accessibility policy statement. The DOJ has also supported students who have sought captioning for online video content as part of a learning platform (*18*).

In Statements of Interest (SOIs) in several recent cases, the DOJ affirmed that Title III covers websites and that the court should not wait for web regulations to be finalized. Moreover, the DOJ stated that public accommodations must make their video content accessible to all people with disabilities. The SOIs assume that the ADA had preexisting coverage of websites. Public accommodations hold an obligation to offer effective communication to ensure equal access to online services.

These various efforts by OCR and DOJ indicate that it is a best practice to make electronic and online content accessible. This includes electronic documents, online courses, and the learning management systems used by universities. Despite the absence of official standards, both agencies have mandated web accessibility for post-secondary institutions.

Physical Access Requirements in Brief

Section 504 and the ADA require that programs and facilities be made accessible to individuals with disabilities. The current enforceable standard for physical access under Title II and III of the ADA is the 2010 ADA Standards for Accessible Design. Under Section 504, each federal enforcement agency determines their standards for physical accessibility. The Department of Education that enforces Section 504 has also determined the 2010 ADA Standards for Accessible Design will apply to post-secondary educational institutions. These standards establish various physical requirements that impact entry and egress for facilities, the design of spaces, the arrangement of tables and equipment, and other elements in the environment. All new construction and alterations as of March 15, 2012 must comply with these standards. Buildings constructed or altered from 1991 to 2012 were required to comply with the 1991 Standards for Accessible Design or the Uniform Federal Accessibility Code depending on the type of entity.

Post-secondary institutions cannot impose fees or charges on students for costs associated with meeting their obligations under federal law. If a reasonable modification or an auxiliary aid or service is denied to a student, it should be based on fact, and not unsubstantiated or stereotypical assumptions about students with disabilities. Determination of disability and appropriate accommodation should be made by qualified professionals in the Office of Disability Services in conjunction with academic professionals and the student with a disability. The ADA protects qualified individuals with a disability who exercise their rights under the ADA, and those who may assist them in that process from retaliation. It also protects people from coercion as they seek to exercise those rights.

Enforcement

Section 504 is enforced by the U.S. Department of Education's OCR. An individual can file a discrimination complaint and OCR may initiate a "compliance review" of a covered entity. OCR's pursue complaints in a variety of areas including: admissions, recruitment, financial aid, academic programs, student treatment and services, counseling and guidance, classroom assignment, grading, athletics, housing, and employment.

In most post-secondary institutions, Title II is enforced by the Department of Education's Office of Civil Rights. Schools related to medicine, dentistry, nursing and other health related fields Title II is enforced by the Department of Health and Human Services (HHS). Title III is enforced by the DOJ. In some cases, where a post-secondary institution is covered by both Title III of the ADA and Section 504 of the Rehabilitation Act, they may defer to the OCR or to HHS. Under Title III, private parties may also bring lawsuits to obtain court orders to stop discrimination. While no monetary awards are allowed in this case, reasonable attorney's fees may be awarded. In some instances, where a pattern or practice of discrimination is alleged, suits may be brought by the attorney general. Under which circumstances, monetary damages and civil penalties may be awarded within limits.

Conclusion

Section 504 of the Rehabilitation Act and the ADA touch deeply on the post-secondary education setting. They require modifications in policies and procedures; they ensure effective communication for individuals with sensory impairments; they require auxiliary aids or services be provided; and they mandate that post-secondary institutions reasonably modify their policies, practices and procedures in order to ensure full access for qualified individuals with disabilities. The intent of these laws is to promote equality of access and equality of opportunity for a group of individuals that historically have been excluded from society. In the following chapters, we shall explore some of these requirements as they relate to these individuals.

Upon entering the post-secondary environment, students may not fully realize what their rights and responsibilities are and subsequently need to adjust to their new situations. Each student, their circumstance, and their disability will be unique. While students with disabilities learn their rights, institutions must learn about the individual. Each person's disability and any educational needs arising from that will be distinct. Even when two students appear to have the same type of disability, low vision for example, there are a myriad of causes for those conditions and each person's experience and ability can vary greatly. Colleges and universities must engage the individual to truly determine what accommodation is necessary and appropriate, which modification in the way a class is conducted must be done, and which solution will serve best to promote the equality intended by Section 504 and the ADA. As the DOJ points out, the "ADA requires equality of opportunity, but does not guarantee equality of results" (*12*). All students must demonstrate mastery in the degree program in order to be successful.

References

1. Schur, L.; Kruse, D.BlanckP. Corporate culture and the employment of persons with disabilities. *Behav. Sci. Law* **2005**, *23*, 3–20; DOI: 10.1002/bsl.624.
2. Yin, M.; Shaewitz, D.; Megra, M. *An uneven playing field: The lack of equal pay for people with disabilities*; American Institute for Research: Washington, DC, 2014. http://www.air.org/resource/uneven-playing-field-lack-equal-pay-people-disabilities (accessed November 16, 2017).
3. Hutcheon, E. J.; Wolbring, G. Voices of "disabled" post secondary students: Examining higher education "disability" policy using an ableism lens. *J. Divers. Higher Educ.* **2012**, *5*, 39–49.
4. Mayerson, A. The History of the Americans with Disabilities Act; DREDF; 1992.https://dredf.org/news/publications/the-history-of-the-ada/ (accessed November 16, 2017).
5. Disability Rights Education & Defense Fund (DREDF). Section 504 of the Rehabilitation Act of 1973. https://dredf.org/legal-advocacy/laws/section-504-of-the-rehabilitation-act-of-1973/ (accessed November 16, 2017).

6. Legal Information Institute. Pub. L. No. 93-112, 87 Stat. 394 (Sept. 26, 1973), codified at 29 U.S.C. § 701 et seq. https://www.law.cornell.edu/uscode/text/29/701 (accessed November 16, 2017).
7. Legal Information Institute. 29 U.S. Code § 794 - Nondiscrimination Under Federal Grants and Programs. https://www.law.cornell.edu/uscode/text/29/794 (accessed November 16, 2017).
8. U.S. Department of Education. 34 CFR 104.3(j)(2)(i) Nondiscrimination on the Basis of Handicap in Programs or Activities Receiving Federal Financial Assistance. https://www2.ed.gov/policy/rights/reg/ocr/edlite-34cfr104.html (accessed November 16, 2017).
9. U.S. Department of Education. Protecting Students with Disabilities https://www2.ed.gov/about/offices/list/ocr/504faq.html (accessed November 16, 2017).
10. U.S. Government Publishing Office. 34 CFR 104.44 Academic Adjustments. https://www.gpo.gov/fdsys/granule/CFR-2010-title34-vol1/CFR-2010-title34-vol1-sec104-44 (accessed November 16, 2017).
11. Kentucky Personnel, n.d. ADA and Postsecondary Education: Title II. https://personnel.ky.gov/OER%20Documents/ADAPostsecondaryEducation.pdf (accessed November 16, 2017).
12. U.S. Department of Justice, n.d. The Americans with Disabilities Act: Title II technical assistance manual covering state and local government programs and services. https://www.ada.gov/taman2.html (accessed November 16, 2017).
13. U.S. Department of Education, n.d. Americans with Disabilities Act: ADA Title III technical assistance manual covering public accommodations and commercial facilities. https://www.ada.gov/taman3.html (accessed November 16, 2017).
14. World Wide Web Consortia. www.w3.org (accessed November 16, 2017).
15. Youngstown State University Resolution Agreement OCR Compliance Review #15-13-6002, Department of Education Office for Civil Rights, 2014. http://www2.ed.gov/documents/press-releases/youngstown-state-university-agreement.pdf (accessed February 3, 2018).
16. Settlement Agreement Between the United States of America, Louisiana TECH University, and the Board of Supervisors for the University of Louisiana System under the Americans with Disabilities Act DJ #204-33-116, Department of Justice Civil Rights Division, 2013. https://www.ada.gov/louisiana-tech.htm (accessed January 3, 2018).
17. Consent Decree: Aleeha Dudley and the United States v. Miami University, et al, Case No.: 1:14-cv-38, Dlott. Susan J. Judge, 2016. https://www.ada.gov/miami_university_cd.html (accessed January 3, 2018).
18. United States of America Statement of Interest in National Association of the Deaf (NAD) V Harvard University. LYNCH. LORETTA E. Attorney General for the United States, 2015, http://www.ada.gov/briefs/harvard_soi.pdf, and United States of America Statement of Interest in NAD V Massachusetts Institute of Technology, LYNCH. LORETTA E. Attorney General for the United States, 2015. http://www.ada.gov/briefs/mit_soi.pdf (accessed January 3, 2018).

Chapter 3

Assessing the Laboratory Environment

Jennifer Perry[*,1] **and Janet Baum**[2]

[1]**Yang-Tan Institute on Employment and Disability, School of Industrial and Labor Relations, Northeast ADA Center, Cornell University, 201 Dolgen Hall, Ithaca, New York 14853, United States**
[2]**T.H. Chan School of Public Health, Harvard University, Executive and Continuing Professional Education, 677 Huntington Avenue, Boston, Massachusetts 02115, United States**
[*]**E-mail: jlp359@cornell.edu**

A laboratory that is accessible will have certain features built into the room that allows access to persons with a wide variety of disabilities. Achieving accessibility in the physical laboratory environment requires attention to both federally mandated standards for access under the Americans with Disabilities Act (ADA) and applicable building code provisions, as well as consideration of unique safety issues that arise in laboratories. This chapter will review accessible design requirements applicable to laboratories, along with discussion of safety issues and best practices for assessing whether a lab is accessible, usable, and safe for end users who have disabilities.

Architectural Considerations, Laboratory Access

There are many ways to meet accessibility requirements. It is important to have an experienced lab planner or architect/engineer who is versed in all building codes and accessibility requirements when designing a lab. The following is a summary of the major "Building Blocks" for accessibility in the 2010 ADA Standards for Accessible Design to be applied in laboratory settings. The 2010 ADA Standards for Accessible Design (*1*), published by the United States Department of Justice (DOJ), are publicly available at: https://www.ada.gov/2010ADAstandards_index.htm.

The United States Access Board, which develops the ADA Standards for accessibility, has a publicly available "Guide to the ADA Standards" that includes the illustrative figures included throughout this chapter (*2*). This guide can be accessed at: https://www.access-board.gov/guidelines-and-standards/buildings-and-sites/about-the-ada-standards/guide-to-the-ada-standards.

While this chapter reviews mandatory architectural components required by the ADA, "Universal Design", an emerging area in architectural design, is defined as the design of products and facilities that can be used by all people, to the greatest extent possible, at low or no extra cost and without the need for adaptation or specialized design. Universal Design benefits people of all ages and abilities and often exceeds the minimum requirements for accessibility detailed below. This chapter reviews the ADA's mandatory minimum requirements for accessibility and focuses upon the direct benefits of ADA that reduce risks and improve health and safety for laboratory workers with disabilities.

Accessible Route

An accessible route allows someone with a disability to approach, enter and use a building or facility. If these routes are not wide enough, or if they have a running or cross slope that exceeds the standards described below, access for people with disabilities may not be possible. For example, cross slopes, the slope that runs perpendicular to the running slope of the ground surface, that exceed two percent can create uneven surfaces that make wheelchair maneuvering difficult because not all wheels rest evenly on the surface, and can cause someone to "tip" or make navigation too difficult.

Elements of accessible routes include walking surfaces, doorways, ramps, elevators, and, where permitted, platform lifts. An accessible route is defined in the ADA Standards as a continuous unobstructed path connecting all accessible elements and spaces of a building or facility. Interior accessible routes may include corridors, floors, ramps, elevators, lifts, and clear floor space at fixtures. The elements of an accessible route are described below.

Elements of an Accessible Route include:

- 80 inches (2.0 m) minimum clear head room;
- 36 inches (91.4 cm) minimum clear width (can be reduced to 32 inches (81.3 cm) minimum in certain circumstances, for a distance not to exceed 24 inches (61.0 cn), i.e. accessible door clear width is permitted to be 32 inches minimum.) See Figure 1.

Passing space is at least 5 feet minimum (1.5 m) x 5 feet minimum provided every 200 feet (61 m), if the route is less than 60 inches (1.5 m) wide, as in Figure 2.

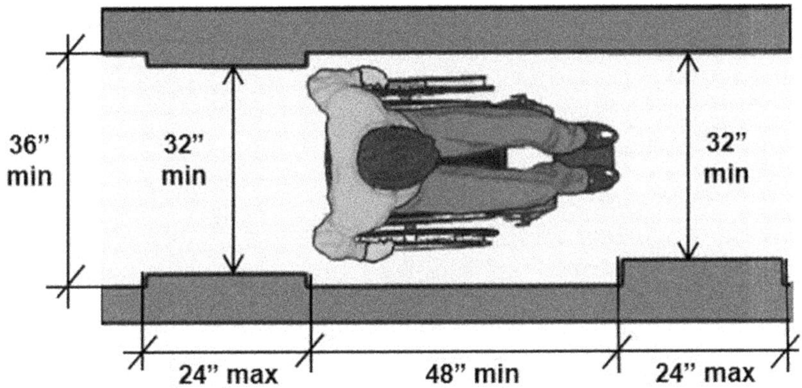

Figure 1. Accessible route. Reprinted from reference (2).

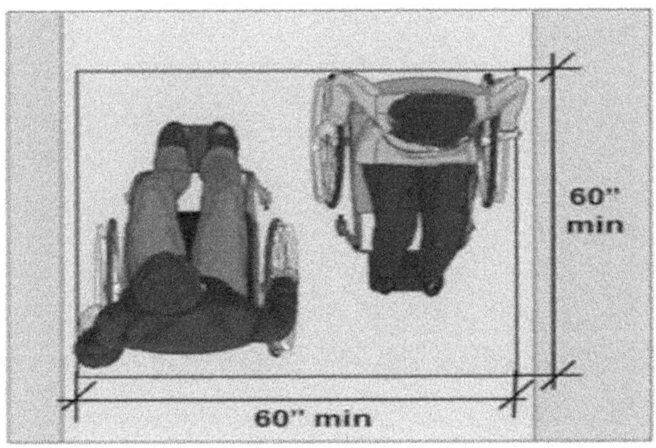

Figure 2. Accessible route passing space. Reprinted from reference (2).

Firm, Stable, Slip Resistant Ground Surfaces

The ADA Standards do not mandate what material must be used for accessible routes, but whatever material is used must meet the performance standard of being stable, firm and slip resistant. Running slope may not exceed 5% (1:20) for walking surfaces; cross slope may not exceed 1:48 (2%). Cross slope is the slope of the surface perpendicular to the direction of travel. Changes in level between ¼ inch (0.64 cm) and ½ inch (1.27 cm) must be beveled with a slope no greater than 1:2. Figures 3 and 4 illustrate this rise.

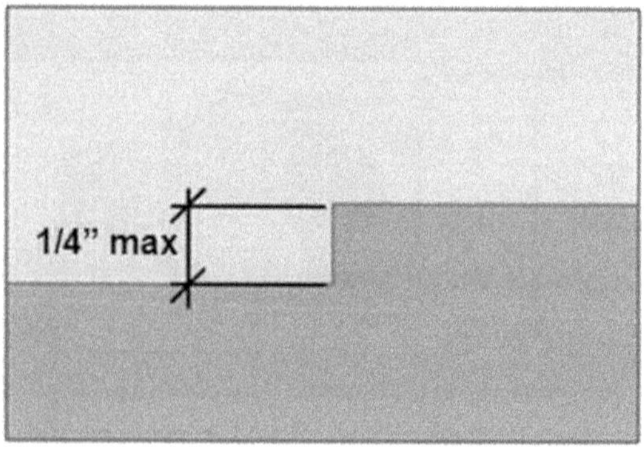

Figure 3. Vertical change in level. Reprinted from reference (2).

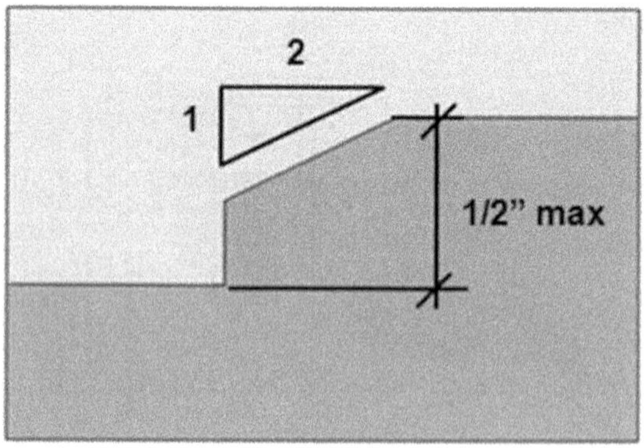

Figure 4. Beveled change in level. Reprinted from reference (2).

Carpeting must be securely attached, with a maximum pile thickness of ½ inch (1.3 cm). Exposed edges must be attached to the ground in order to avoid tripping hazards.

Gratings must have spaces no greater than ½ inch (1.3 cm) wide in one direction. Elongated openings, like those of most grates, must be oriented so that the long dimension is perpendicular to the dominant travel direction. If this is not done, wheelchair casters, canes or crutch tips can get wedged into wider openings. See Figure 5.

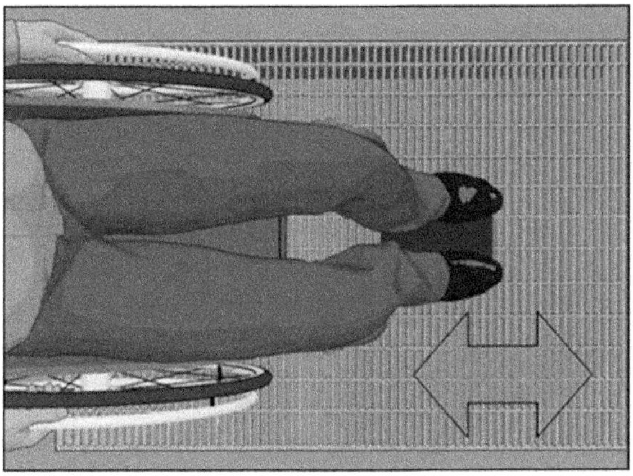

Figure 5. Gratings. Reprinted from reference (2).

Ramps and curb ramps are required along accessible routes to span changes in level greater than ½ inch (1.3 cm). Elevators and, under certain specified conditions, platform lifts, can be used as an alternative. Portions of accessible routes with running slopes steeper than 5% must be treated as ramps, not as walking surfaces.

Where ramps are part of an accessible route, the technical requirements are as follows. See Figure 6 for a complete detail of ADA ramp specifications. These include:

- Maximum running slope of 1:12 (8.33%);
- Maximum cross slope is 1:48 (2%);
- The height of runs is limited to 30 inches (76.2 cm) maximum;
- Ramp runs must have a clear width of 36 inches (91.4 cm) minimum (measured between handrails where provided). In addition, within Employee Work Areas (areas used ONLY by employees and ONLY for work), common use **circulation paths** must be accessible in work areas 1,000 sq. ft. (92.9 sm) or more in size;
- Level landings are required at the top and bottom of each run;
- Handrails are required on both sides of ramps with a rise greater than 6 inches (15.2 cm).

Ramps on **circulation paths** must comply exceptions:

- The clear width can be reduced below 36 inches (91.4 cm) by work area equipment where it is essential to the work being performed; and
- Handrails can be installed after construction, as needed. However, ramps must be sized so that the minimum clear width is maintained.

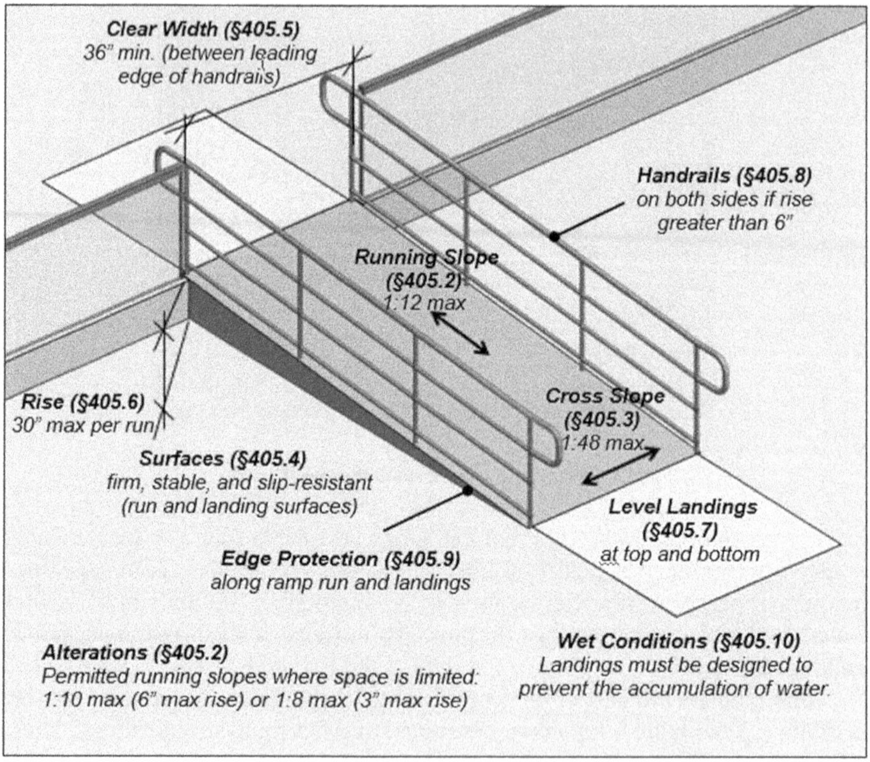

Figure 6. Ramp details. Reprinted from reference (2).

Edge protection along ramp runs and landings must be provided to keep wheelchair casters and crutch tips on the surface of the ramp. Edge protection can include extending the ground surface of the ramp run 12 inches (30.5 cm) minimum beyond the inside face of the handrail. The alternative is to have a curb or barrier that can run along the bottom of the ramp surface which prevents the passage of a 4 inch (10.2 cm) sphere where any portion of the sphere is within 4 inches of the finish floor or ground surface. See Figures 7 and 8 below.

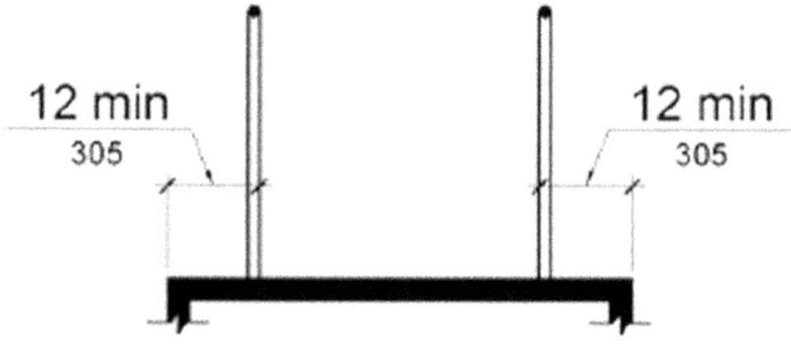

Figure 7. Ramp surface edge protection. Reprinted from reference (2).

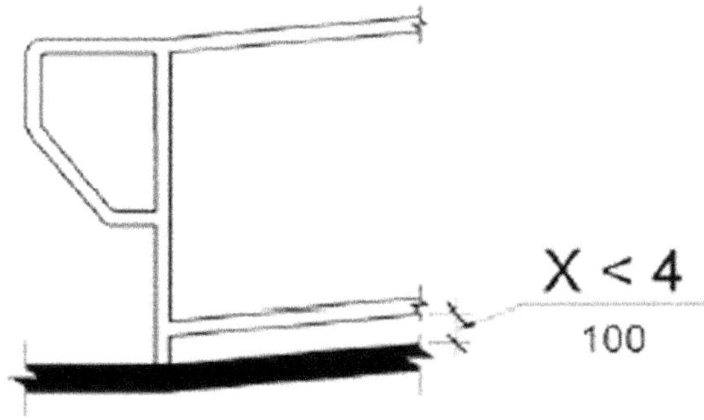

Figure 8. Ramp surface edge protection with curb or barrier option. Reprinted from reference (2).

Intermediate landings between runs must be at least 60 inches (1.5 m) wide and 60 inches long where ramps change direction (any change from linear). The landings cannot have any obstructions within this 60 inch (1.5 m) minimum x 60 inch minimum space. See Figure 9.

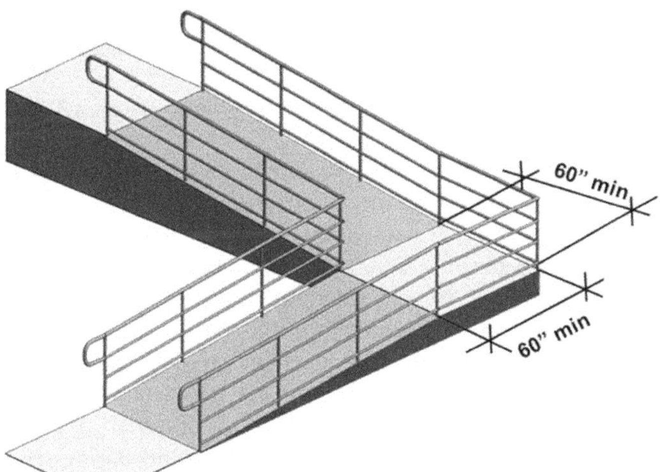

Figure 9. Landing where ramps change direction. Reprinted from reference (2).

Accessible Doors

Doors and gates along an accessible route (doors you pass through) must be accessible to people with disabilities. Many people mistakenly assume that the ADA requires doors to be automatic or power assisted to be accessible, but this is not the case. There are other technical requirements that apply to doors that make them accessible. At least one accessible door or gate serving each accessible room, space, and entrance must comply with the ADA Standards. Below are the requirements for accessible doors.

Doors used for entry or exit in laboratories have major safety considerations. A minimum of 32 inches of clear width is required. This is measured from the stop to the face of doors or gates open to a 90 degree position (or to the leading edge of sliding or folding doors). No projection into the clear width is permitted below 34 inches (86.4 cm). At least one active leaf where double doors are provided must meet this 32 inches (81.3 cm) minimum clear width requirement, as illustrated in Figure 10.

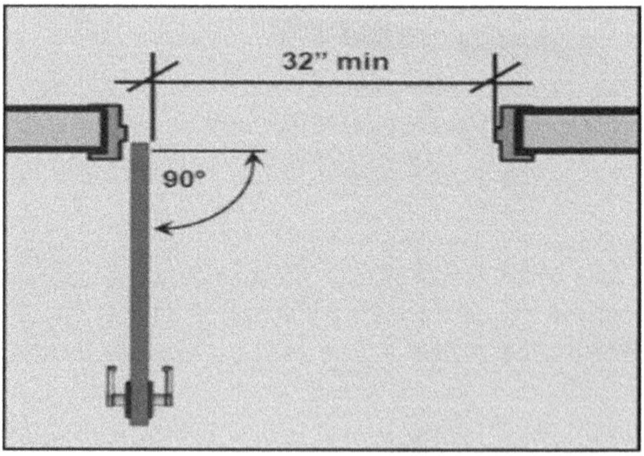

Figure 10. Door width. Reprinted from reference (2).

Wider, single-leaf doors, between 36 and 42 inches (91 and 107 cm) benefit persons using standard wheelchairs, wide-wheelchairs now commonly used, and other mobility devices to pass through more comfortably. Wider doors allow workers who are pushing laboratory carts, hand-trucks, mobile scientific or housekeeping equipment, to pass through the door more safely, reducing the possibility of damage or an accidental spill.

Electric-powered, automatic door openers provide more convenience to workers with mobility limitations. Automatic door-opener installations that comply with ADA Guidelines may reduce ADA required clearances on one or both sides of the door (*1*). This is particularly useful during renovations of older laboratory buildings where out-swinging entry doors are often recessed in narrow alcoves that lack ADA compliant clearances. With automatic door opener installations, appropriate signage and location of the activation button must meet ADA standards.

Double doors facilitate safety and ease of handling bulky items when they are routinely brought in and out of laboratories. The door's active-leaf must be at least 36-inch (91 cm) minimum width for workers using wheelchairs (*3*). The inactive-leaf may be a width most suitable for providing clearance for bulky items, 12 to 36 inches (30.5 to 91 cm). Door hardware selected to close and keep the inactive-leaf shut affects convenience for workers using that door. Hardware also affects the door security.

Another safety consideration is that emergency egress laboratory doors must swing outward in the direction of travel (*4*). Within a laboratory space, accidental explosions, fires, or release of large amounts of compressed or cryogenic gases can cause over-pressurization of the lab space. When occupants need to evacuate, exit doors that swing inward to the laboratory can become a hazard. In such emergencies, out-swinging exit doors allow quick egress for workers heading to the same door. However, in the event that the laboratory building must be evacuated, out-swinging doors cannot impinge upon corridor widths, required by all national building codes, that allow safe emergency egress of occupants exiting via the corridors. Out-swinging doors are either shielded by provision of entry alcoves at doors recessed from the corridor, or by designing corridors to meet code requirements that compensate for area that open doors occupy.

There are several requirements for interior doors, as illustrated in Figure 11. If doors have closers, it must take at least 5 seconds for the door to close between the 90 degrees and 12 degrees position. This prevents mobility devices from becoming trapped between doors that close too quickly.

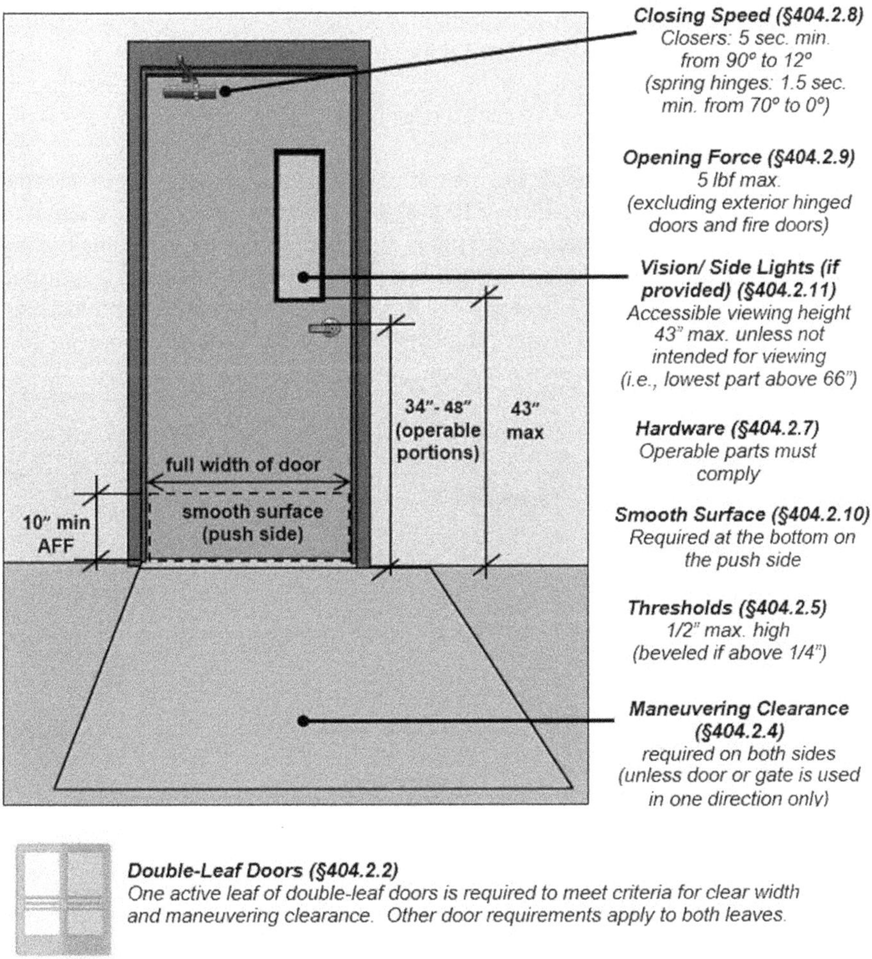

Figure 11. ADA door requirements. Reprinted from reference (2).

Interior doors cannot require more than 5 pounds of force (2.27 kg) to operate (push/pull). The maximum of 5 pounds of force does not apply to exterior hinged doors or fire doors. Note that many interior doors with closers can be improved by adjusting the closer on the door to require less force when pushing/pulling.

If vision or sidelights are provided at doors, the lowest edge of the vision/sidelight must be located no higher than 43 inches (1.1 m) above the ground to permit viewing by people with disabilities.

Door hardware must be accessible and not require any tight grasping, pinching or twisting of the wrist to operate. Lever style hardware is an example of accessible door hardware. It is advisable to avoid hardware that requires hand or finger dexterity, fine motor movement, or the use of both hands for operation. For example, round door knobs are not accessible as they require twisting of the wrist.

On doors where you "push" the door to open, the bottom 10 inches (25.4 cm) of the door surface must be smooth so that someone pushing against the door with the footrests of a wheelchair or with a walker can push against the door.

The height of door thresholds is limited to ½ inch (1.3 cm) in new construction. The edge must be beveled with a 1:2 maximum slope above a height of ¼ inch (0.64 cm). A maximum height of ¾ inch (1.9 cm) is permitted for existing or altered thresholds if they have a beveled edge on each side with a slope not steeper than 1:2.

Doors must have maneuvering clearance, which is defined as the space that permits someone using a wheelchair or mobility device to move close enough to the door to push/pull the door open. This is illustrated in Figure 12. Maneuvering clearances are determined based on the direction of approach, swing of doors, and in some cases the presence of a closer or latch. They are required on both sides of doors or gates except at those that can be used in one direction only.

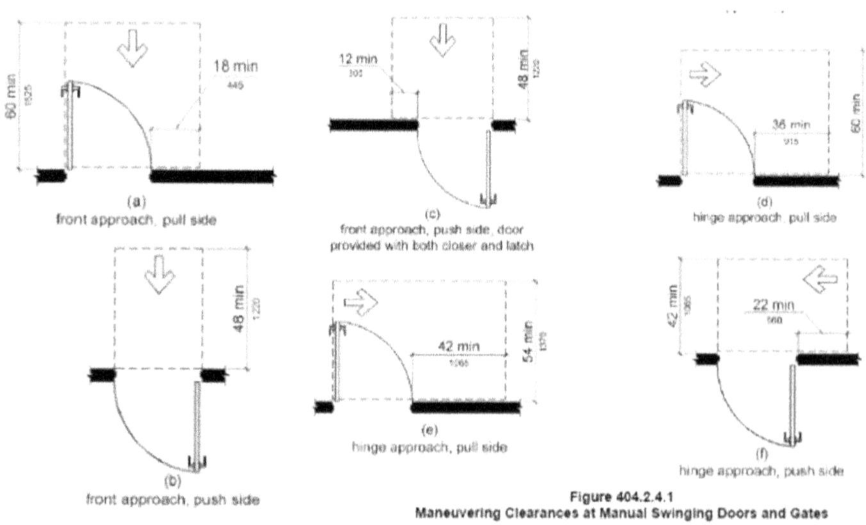

Figure 12. Doors swings, clearances for wheelchair approaches. Reprinted from reference (1).

Accessible Signage

Within a laboratory facility, it is important that accessible signage is provided. Particularly for people with visual disabilities, labeling what are referred to as "permanent rooms and spaces" with accessible signage will also be of use to those may not be familiar with the facility. Permanent rooms and spaces are those that are not likely to change in use over time, such as labs, toilet rooms, janitor's closets, break rooms, room numbers, and exit signs. These should have the following features:

- Text characters must contrast with their backgrounds (i.e. black on white or white on black);
- Text characters must be raised (tactile);
- Braille must be provided;
- Signs must be mounted on the wall on the latch side of the door;
- The baseline of the lowest sign character must be at least 48 inches above the floor and the baseline of the highest character must be no more than 60 inches above the floor.

Figure 13 illustrates a sign designating a permanent room. If the sign is at double doors with one active leaf, the sign should be on the inactive leaf; if both leaves are active, the sign should be on the wall to the right of the right leaf.

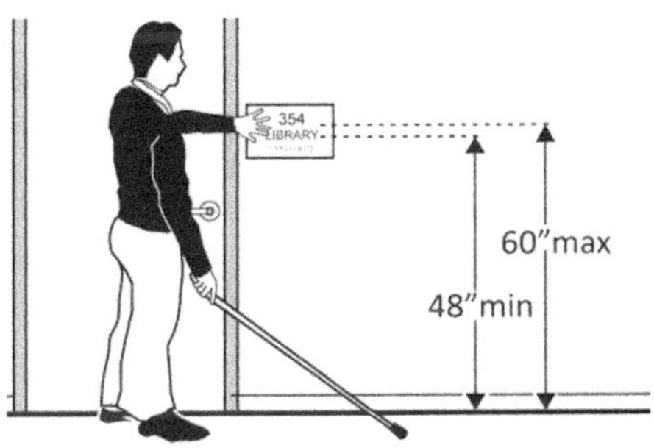

Figure 13. Door sign height and tactile reading. Reprinted from reference (4).

Other signs that provide directions or information about a facility, should have the features bulleted below. Examples of directional or informational signs include those indicating evacuation procedures, directions to accessible toilet rooms, and "No Smoking" signs. Raised characters and Braille are not required at directional and informational signs.

- Text characters must contrast with their backgrounds (i.e. black on white or white on black);
- Signs must be mounted so that the characters are at least 40 inches (1.0 m) above the floor, as in Figure 14;
- The text characters contrast with their background and the sign is at 40 inches (1.0 m) from the characters to the floor.

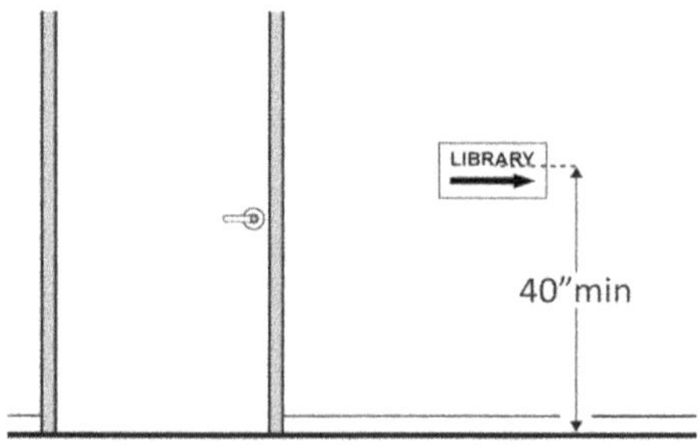

Figure 14. Directional sign at proper height. Reprinted from reference (6).

Chapter 6 of this book discusses goals, techniques, and guidelines to provide accommodations for blind persons or with limited vision. There are a wide variety of visual impairments. Trained professionals, as described throughout this book, must be available to work with each individual to provide facilities' accommodations and administrative support for their employees with vision disabilities. Considerations that impact assessments and facility accommodations that may be needed for a range of vision impairments, including color-blindness.

Signage that has large, clean, simple fonts, high-contrast lettering and symbols improve visibility. A trend of signage in shades of gray lettering on gray backgrounds, that many architects and signage consultants favor for its aesthetic value, make it difficult for many persons to see clearly or accurately. Black letters on black backgrounds are even harder to see.

Signs with complicated or unfamiliar fonts can be eye-catching in advertisements and on store signs, but are not appropriate when used for critical safety and way-finding signs found in scientific laboratory settings. Examples of critical signage with simple, easy to read, large fonts, include those shown in Figures 15A and 15B. Signs should also have OSHA-approved colors, sizes, and formats for safety information (*4*, *5*).

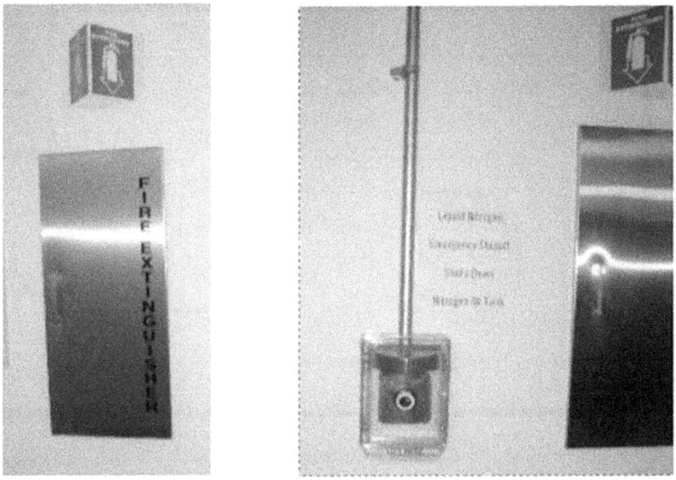

(A) (B)

Figure 15. Signs and tags for safety devices. (A) Fire extinguisher and cabinet. (b) Sign and liquid nitrogen emergency shut-off. Published with permission from Janet Baum, author.

Laboratory Layout

The ADA Standards address clearances and space needed to ensure that the laboratory room is usable by individuals with disabilities. The following accessible features addressed here include clear floor or ground space and turning space; knee and toe clearance; accessible reach ranges and operable parts. These elements serve as the "foundation" that ensures access to specific elements that are often critical in laboratories, such as lab desks and sinks.

Persons working in laboratories need to access many areas, such as shared equipment zones, laboratory chemical hoods, chemical storage and hazardous waste disposal zones, sinks, and emergency response equipment. Being able to make 90 degree and 180 degree turns is essential in work areas that require wheelchair users to pull up to and then back out from workstations. The goal for the laboratory design is to provide safe passage in aisles from every zone of the laboratory that lead to laboratory exits without causing obstruction, particularly during emergency egress conditions. Wheelchair turning spaces is addressed in more detail later in this chapter.

The floor plan diagram in Figure 16 illustrates advantages of hazard zoning for workers who use wheelchairs (*5*). Dashed circles show there is adequate clearance for a person in a wheelchair to turn around, allowing workers using wheelchairs to have a choice of direction to move in the lab. Providing a second exit door allows for options for exiting safely in response to emergencies.

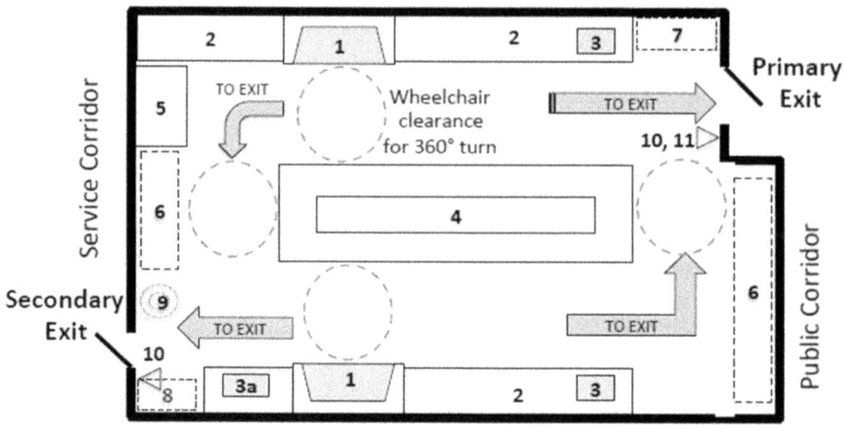

KEY

1 - Chemical hood
2 – Lab bench & wall storage above
3 – Lab sink & drying rack
3a- Lab sink & emergency eye-wash
4 – Lab bench (island) & reagent shelves
5 - Lab bench (mobile) or storage unit
6 – Equipment zone
7 – Lab coat & PPE storage
8 – Hazardous waste storage
9 – Emergency deluge shower
10 – Fire extinguisher (hand held)
11 – Fire blanket & first aid kit

Figure 16. Wheelchair accessibility in a laboratory. Published with permission from Janet Baum, author.

The direction-of-swing and width of doors within laboratory suites that connect to other laboratory spaces, offices, or ancillary functions, need to be evaluated for which options work best for ADA-compliant accessibility and the safety considerations of the laboratory. Under OSHA, as Figure 17 summarizes, an alternate path of egress is required in any high hazard zone (*4*).

> "The safest arrangement for laboratory egress is for [a] required exit door to open into a separate fire zone and for each exit to be located so the pathways within a laboratory or laboratory suite are separated, as far apart as feasible. Thus, when an accident or other emergency makes one laboratory escape pathway impassable, the second can provide an alternative safe route out to another fire-rated building egress pathway" (*6*).

"EXIT ROUTES must be kept free of explosive or highly flammable furnishings or other decorations."
"EXIT ROUTES must be arranged so that employees will not have to travel toward a high hazard area, unless the path of travel is effectively shielded from the high hazard area by suitable partitions or other physical barriers."

Figure 17. OSHA Exit route requirements. Reprinted from reference (4).

Egress provisions are critical for all laboratory workers, but especially those who have functional limitations. These manifest in slow sensory perception for the need to exit immediately or need for increased reaction time to move away from the hazard. In Figure 18 arrows show two paths of egress from this laboratory that contains two chemical hoods (*6*). This diagram assumes that the chemical hood zone is the highest hazard area in the laboratory. In other circumstances, that "highest hazard" may be where hazardous equipment or a hazardous process is located outside a chemical hood or glove box. To achieve an emergency egress advantage, consider an island configuration in lieu of peninsulas, unless exit doors are located at the end of each side of a peninsula bench (*6*).

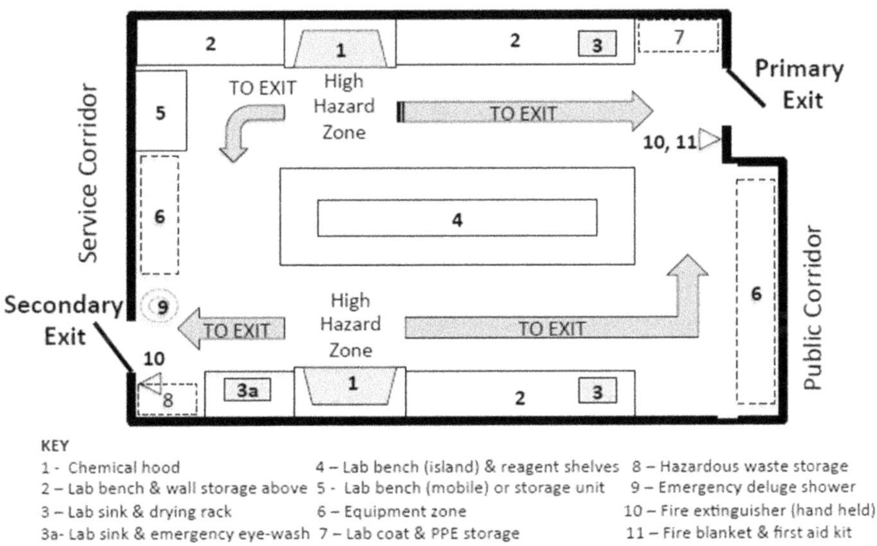

Figure 18. Hazard zoning concept. Published with permission form Janet Baum, author.

Access within a Laboratory

Figure 19 shows detailed requirements for minimum clearances for 90° and 360° turns in wheelchairs. Typical laboratory aisles should be 5-feet (1.5 meters). In addition to meeting ADA requirements for turns, this clearance enables two persons walking to safely pass by each other without physical contact; persons can also safely pass behind workers at the bench without disturbing them. Workers can safely use lab stools or chairs, at knee-spaces under the bench in 5-feet aisles. Additional space in aisles enhances safety throughout the laboratory.

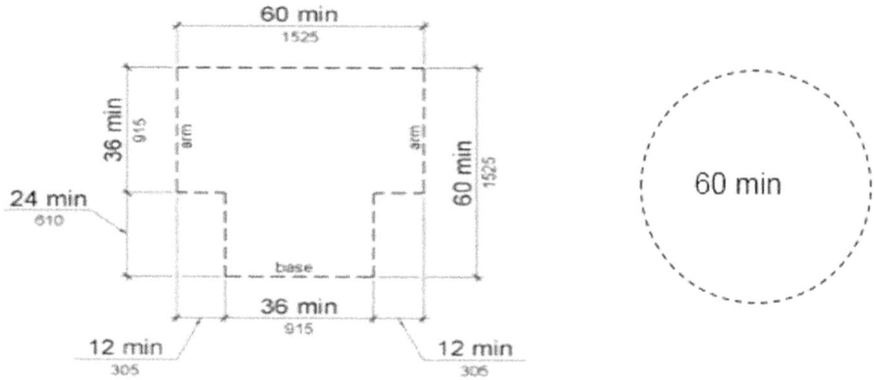

Figure 19. Wheelchair turning space and aisle widths. Reprinted from reference (1).

Recommended clearances in all laboratory aisles should increase to 6-feet (1.83 m) in laboratories that are occupied by two or more wheelchair users. This allows two persons using standard wheelchairs to pass by each other without impeding the other. This width allows persons using wide wheelchairs to pass behind workers at the bench without physical contact. Wider aisles also apply to teaching laboratories where students, can be physically too large/wide to pass by each other (*6*, *7*).

Clear Floor, Ground Space and Turning Space

Clearances are required at accessible elements, fixtures, and controls in laboratory facilities so that people with disabilities, including those who use wheelchairs, can approach and use them. These clearances must be taken into consideration for at least 5% (but not less than one) of laboratory features such as desks, benches, emergency eyewash/shower stations, and chemical hoods. Essentially, at least one of each unique element in the laboratory should be located on an accessible route as discussed previously, and have the space or clearance necessary for someone with a disability to approach and use the element/equipment.

At most elements, clear floor or ground space can be positioned for either a forward or a side approach. For better usability in a laboratory, a forward approach to certain elements, including work surfaces (such as desks), drinking fountains, lavatories, and most sinks is recommended. Forward approach is best these elements because of the need to "roll underneath" of surfaces such as sinks and desks, as opposed to pulling up alongside a desk or sink, which renders these elements less usable. When full access to a desk is not necessary, such as for very short work periods, a side approach to a counter is acceptable. Technical requirements for the clear floor or ground space include:

- Clear floor space minimum size of 30 inches by 48 inches (76 cm by 122 cm) applies whether the space is positioned for forward or side approaches;
- Additional space is required when the space is confined on three sides and is obstructed for more than half the depth, such as when elements are recessed in alcoves. See Figure 20.

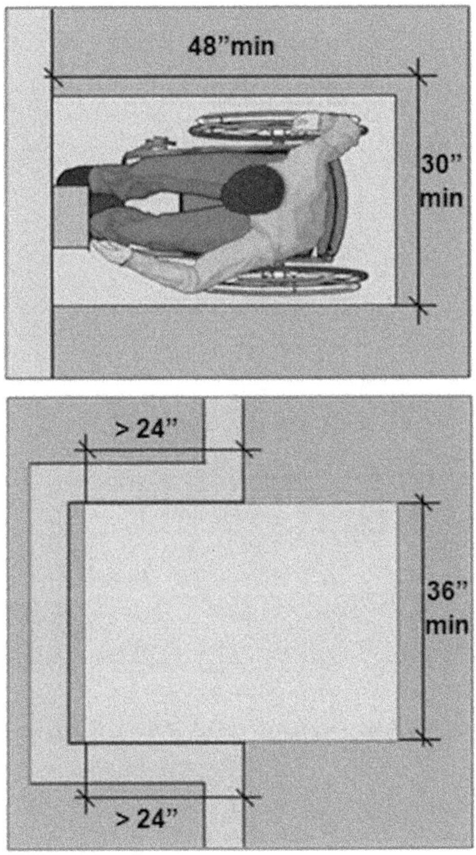

Figure 20. Clear floor space for a forward approach. Reprinted from reference (2).

Knee and Toe Clearance

Knee and toe clearance allows a closer approach to elements and reduces the reach required to operable parts. See Figure 21 for an illustration showing toe clearance (top) and knee clearance (bottom).

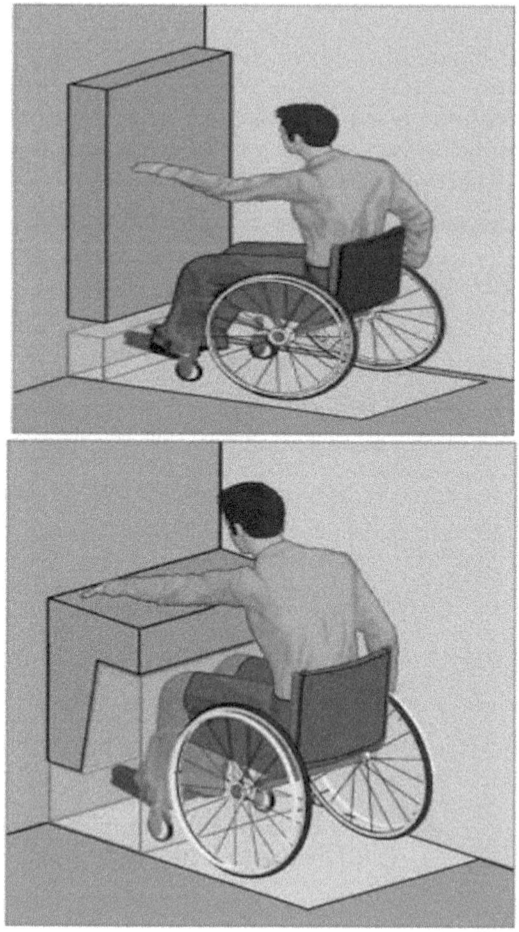

Figure 21. Knee well. Reprinted from reference (2).

Depth of Knee and Toe Clearance

Where knee and toe space is required, it must be at least 17 inches (43.2 cm) deep. In all cases, the minimum depth may be further determined by the required reach to operable parts served by the clear floor space. Knee and toe space is required below drinking fountains, lavatories, and sinks, as well as work surfaces, such as a laboratory desk or sink, so that someone can be positioned with their knees and toes underneath these elements. This is illustrated in Figure 22.

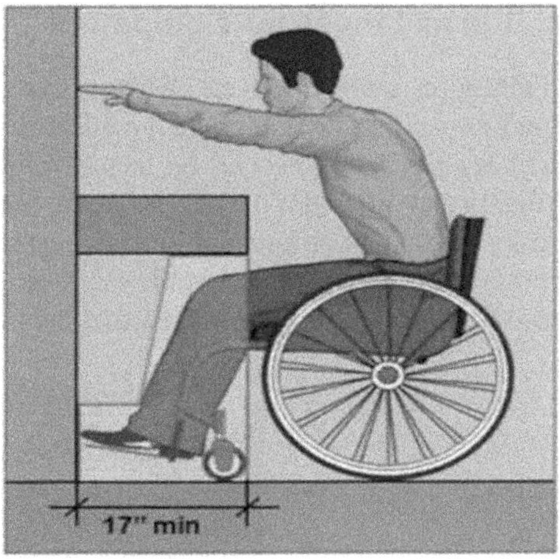

Figure 22. Knee well and reaching. Reprinted from reference (2).

C-Frames and Table-Frame benches can be installed at seated or standing heights. To be ADA compliant, these systems must be installed with the lower position at 32 in (81 cm) to 30 in (76 cm), above a finished floor as illustrated in Figure 23. Lower level countertops also become good locations to place tall equipment or top-loading equipment that is otherwise difficult to reach. Fine motor control and visual access improve when knee-space is provided for workers to sit comfortably at the bench at all counter heights.

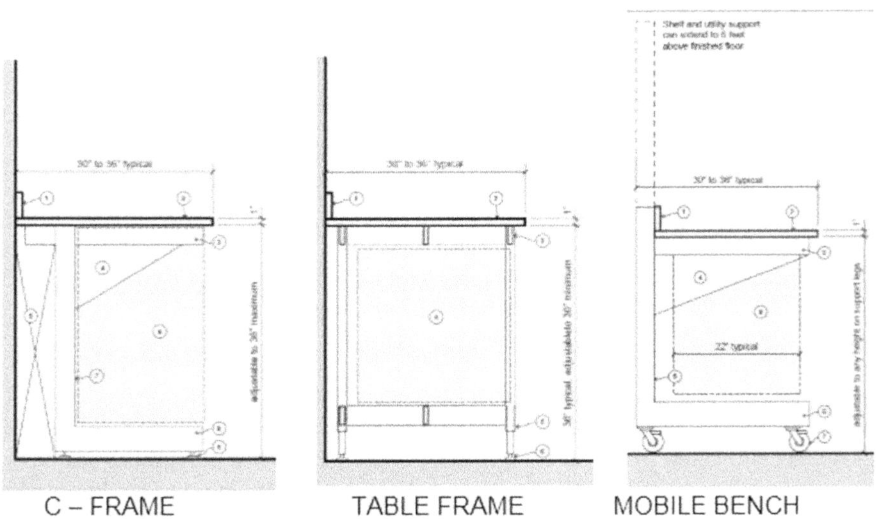

Figure 23. Section diagram of wheelchair accessible lab benches (removable cabinet). Drawing by Janet Baum, author.

Furnishings and Equipment

The ADA Standards apply mainly to those elements that are fixed or built-in. Moveable elements and furnishings are generally not addressed or covered by the ADA Standards. ADA regulations, however, do include requirements that may impact non-fixed elements that fall under the category of a "work surface". These include laboratory desks. Providing an accessible work surface is often achieved through careful selection of at least one accessible lab desk per lab, or through more innovative methods, as shown in Figures 24A and 24B.

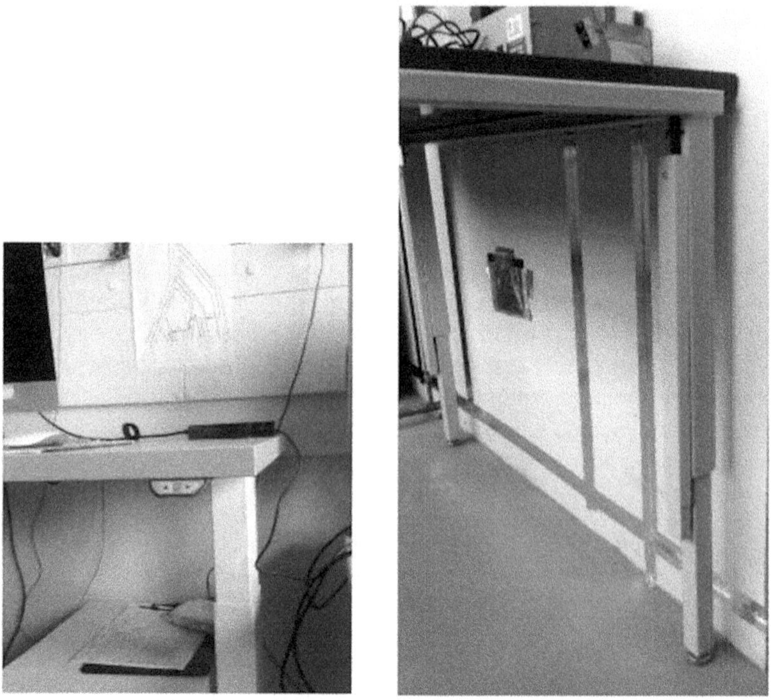

(A) (B)

Figure 24. Adjustable tables. (A) Push button control. (B) Adjustable legs. Published with permission from Ellen Sweet, editor.

Through the use of simple push-button operation that complies with the ADA Standards for accessible operable parts, the height of the lab bench shown in Figure 24A can be lowered to also accommodate people with low stature, those seated or using a mobility device, or standing researchers. In order to meet ADA Standards, side to side clearance that is a minimum 30 inches (76.2 cm) wide must be provided at the desk, or kneehole at an accessible bench (*1*). This width allows users, of standard-size adult-wheel chairs, to safely remove their hands from wheel rims. Desk supports, brackets or legs cannot interfere with this minimum 30 inch (76.2 cm) width. At least 27 inches (68.6 cm) of clearance must be provided underneath the surface of the desk when measured to the finish floor and the height of the desk must be adjustable to a range between 28 inches and 34 inches (71.1 and 86.4 cm) above finish floor. Figure 25A and 25B show accessible chemical hood installations. Figure 25A shows chemical storage cabinets with lockable wheels. This allows for utilization of the space under the hood, but when someone needs to sit and work at the hood they can be removed. Figure 25B shows a chemical hood on an adjustable bench. The ductwork and plumbing have to have flexible connections to allow for adjustability. Utilizing a motorized lift, rather than a hand crank, allows greater accessibility.

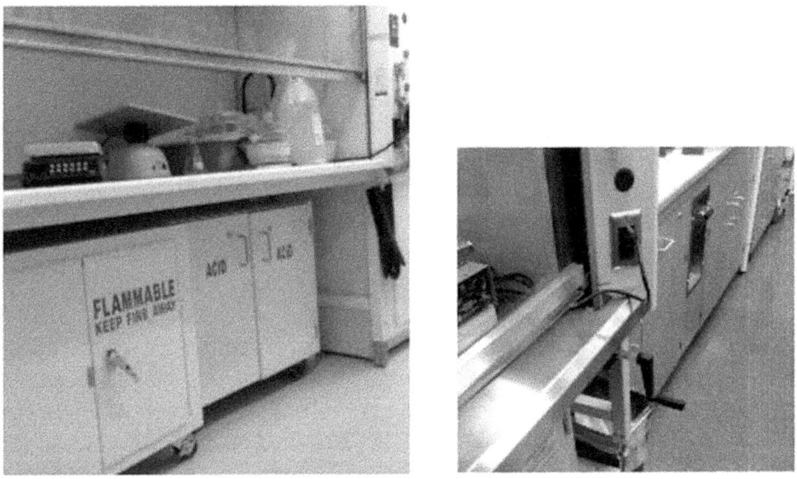

(A) (B)

Figure 25. Examples of Chemical Hood Installation. (A) Flexible installation of chemical storage with lockable wheels. (B) Hand crank to adjust height of chemical hood. Published with permission from Ellen Sweet, editor.

Maintenance of Accessible Features

The ADA Standards address the design, but not the maintenance, of building elements and features. However, it is important that accessible features be properly maintained to ensure working order, except for isolated or temporary interruptions in service due to maintenance or repairs.

Within a lab setting, something as simple as where you place equipment or furniture can eliminate access for people with disabilities if users of the lab are not mindful of the ongoing need to evaluate the space for barriers. With this in mind, some simple things can be done to maintain access in labs include:

- Do not store or place items, equipment, furniture, or other elements in areas where they will reduce the clearance for an accessible route throughout the lab that is at least 36 inches (91.4 cm) wide;
- Do not place safety equipment or other items essential to the lab in areas that are not located on an accessible route;

- Do not block doors or cabinets with elements that will render them unusable by someone with a disability;
- Do not place items, such as cabinets, directly against the frame of doors that must be used for access. This placement can eliminate the clearance needed for someone using a wheelchair. They need to access the door hardware and position themselves beyond the swing of the door.

Obstructed Reach to an Element Depth

At any element, such as a sink or lab desk, the knee and toe space must be deep enough to reach operable parts. Functionally, someone cannot lean forward from a seated position and grasp something that extends beyond their toes, the depth of the knee and toe space beneath an element is limited to 25 inches maximum (63.5 cm) measured from the leading edge of obstructions. See Figure 26 for clarification.

Both knee and toe clearance have been provided at the lab sink shown in Figure 27. This means that there are at least 27 inches (68.9 cm) of clearance underneath the apron of the sink/counter to permit someone using a wheelchair to roll underneath of the sink. The maximum height to the top of the sink and surrounding counter is 34 inches (86.4 cm) above finish floor. Hot water and drain pipes should always be insulated or otherwise configured so as to protect from contact by users. A minimum of 30 inches (76.2 cm) of width is required at this area where the knee and toe clearance is provided. Also note the design approach used here placed the operable parts for the sink along the sides of the sink, which increases usability for people with difficulty reaching.

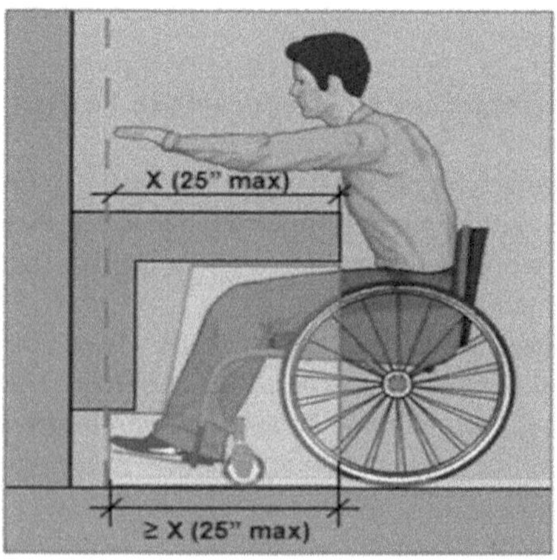

Figure 26. Obstructed reach. Reprinted from reference (2).

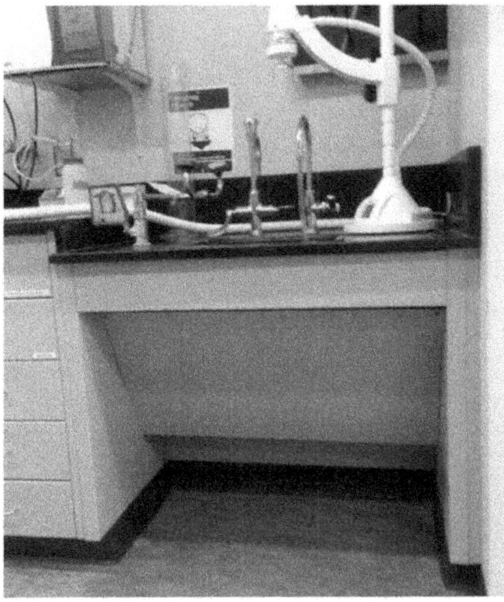

Figure 27. Sink at 34 inches height. Published with permission from Ellen Sweet, editor.

Wheelchair Turning Spaces

Sufficient space is needed to make complete turns, particularly for people that use mobility devices. Positioning to make a 180 degree turn varies by person and the mobility device they use. A common way of turning when using a manual wheelchair is to turn the wheels in opposite directions for a pivoting turn as in Figure 28. Some power chairs also may permit tight circular turns as in Figure 29.

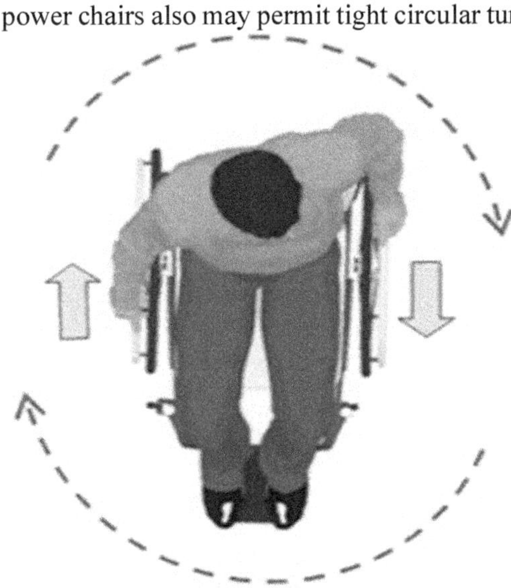

Figure 28. 180 Degree turn (circle) in a manual wheelchair. Reprinted from reference (2).

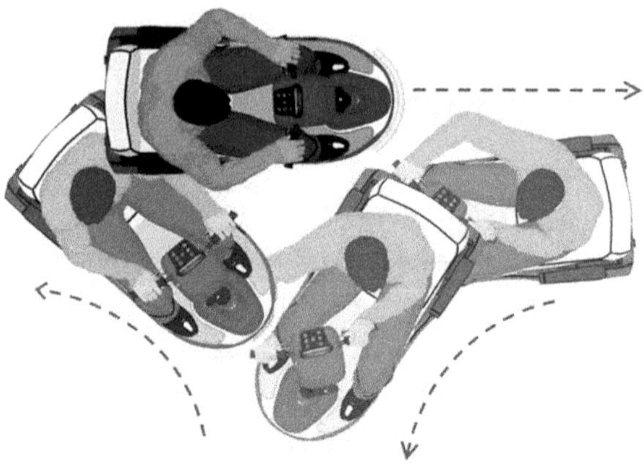

Figure 29. 180 Degree turn in a power chair or scooter. Reprinted from reference (2).

Turning space can be provided in the shape of a circle or a "T". The "T" space is similar to a three-point turn concept. Note that elements with both knee and toe space can overlap a portion of the turning space. See Figure 30.

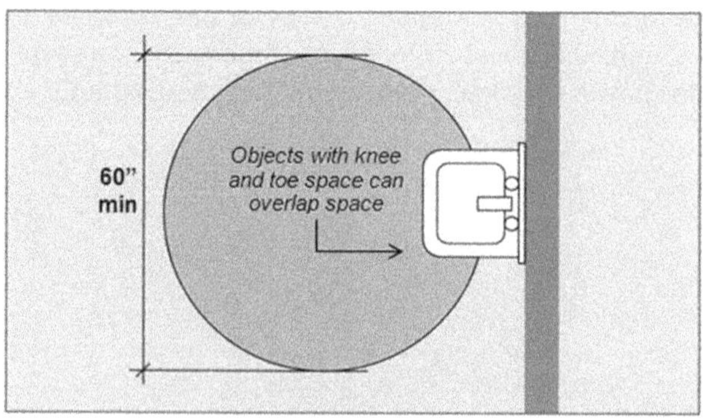

Figure 30. Circular wheelchair turning space. Reprinted from reference (2).

The T-shaped turning space can be arranged for approach on any segment (stem or either arm) in order to make a turn. See Figure 31.

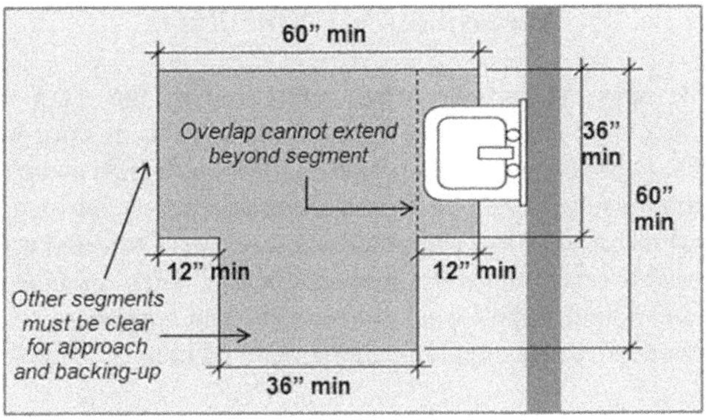

Figure 31. T-Shaped turning space. Reprinted from reference (2).

Along accessible routes, turning space may be necessary when making tight turns around objects that are less than 48 inches (1.2 m) wide, such as between shelving or equipment placed close together or stacks in a library. In these cases, as in Figure 32 where maneuvering is tight, at least 60 inches (1.5 m) of space is needed at the top of the turn (as opposed to 36 inches (91.4 cm) minimum which is typically required for an accessible route). Finally, the turning space can overlap other clearances, including other clear floor spaces, door maneuvering clearances, and fixture clearances.

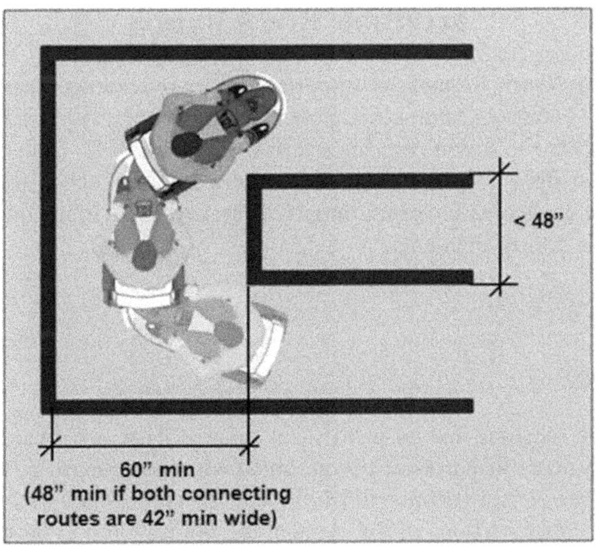

Figure 32. Turning space. Reprinted from reference (2).

Accessible Operable Parts

Accessible operable parts are often referenced by the ADA standards for building elements and certain types of equipment, such as door hardware and faucets at sinks. Other examples of operable parts include light switches, electrical receptacles, thermostats, alarm pulls and automatic door controls (if provided). Floor electrical receptacles and receptacles that serve a dedicated use are exempt from the accessible reach range requirements below. These include phone jacks, data ports, network and audio-visual connections, and appliances.

Key components of accessible operable parts include:

- No tight grasping, pinching or twisting of the wrist to operate;
- Operation with one hand;
- No more than 5 pounds (2.27 kg) of force for operation;
- Located within accessible reach ranges.

When specifying operable parts, note that standard U-shaped pulls and lever-shaped handles are acceptable. Stationary knobs with a shape that can be loosely gripped also are acceptable. Knobs that require a full hand grip and turning, including round door knobs, do not comply because they require twisting of the wrist.

Accessible Reach Ranges

Accessible Reach Ranges are important to the overall accessibility of a lab, as elements and controls cannot be usable by many people with disabilities if they are located either too low or too high above the finish floor. As some elements are approached from a side position, and others are approached from a forward position, there are accessible reach range requirements that hinge on the approach to the operable part of elements.

Forward Reach

If you are reaching for an element or control from a forward position, or forward reach, operable parts must be mounted within the range of 15 inches (38.1 cm) to 48 inches (1.22 m) above finish floor when approached in this manner, as in Figure 33. Other portions of the element are not required to be located within this range.

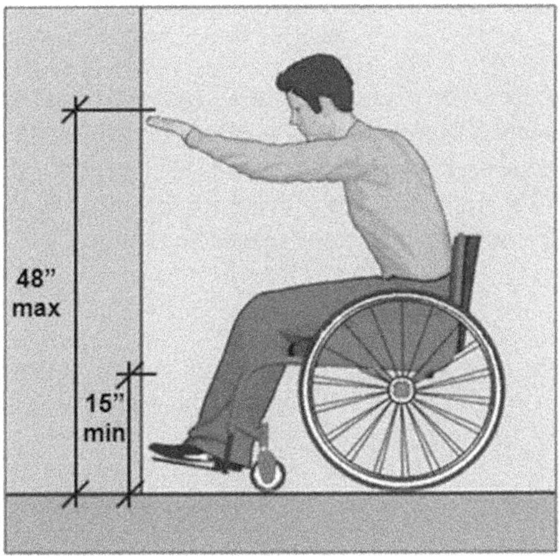

Figure 33. Forward reach. Reprinted from reference (2).

Obstructed High Forward Reach

If an operable part is obstructed, by furniture or equipment for instance, then the maximum high reach of 48 inches (121.9 cm) is reduced to 44 inches (1.12 m) when the depth of the reach over an obstruction exceeds 20 inches (50.8 cm). Knee and toe space must extend the full depth of reach. This is important for operable parts such as faucets that are located at the back of a desk which require reaching over the desk for access. See Figure 34.

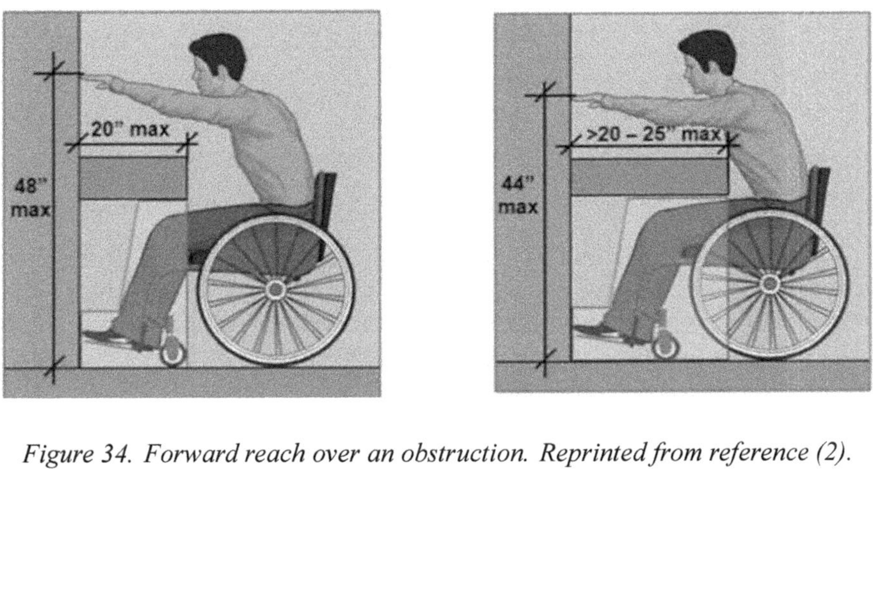

Figure 34. Forward reach over an obstruction. Reprinted from reference (2).

Side Reach

The range for accessible side reach to an operable part, is the same as that for elements approached from a forward position, so the height of operable parts, if unobstructed, must be between 15 inches (38.1 cm) and 48 inches (1.22 m) above finish floor. The maximum reach depth for this range is 10 inches (25.4 cm) measured from the available clear floor space. See Figure 35.

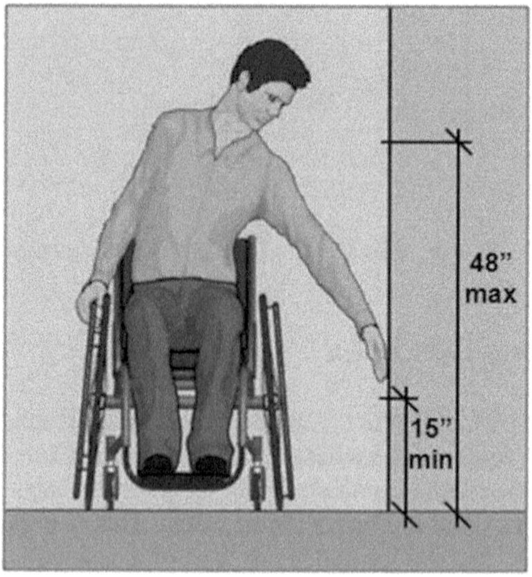

Figure 35. Height of operable parts accessed from an unobstructed side position. Reprinted from reference (2).

Obstructed High Side Reach

If an operable part is obstructed when being reached from the side, by furniture or equipment for instance, then the maximum high reach of 48 inches is reduced to 46 inches (116.8 cm) when the reach over an obstruction is deeper than 10 inches (25.4 cm) (to a maximum of 24 inches (61.0 cm)). Obstructions at side reaches are limited to a height of 34 inches (86.4). See Figures 36 and 37.

Operable parts in corners, such as the end of a counter with casework underneath, are particularly difficult to reach from a side position if seated in a wheelchair. This is because in corners it is not always possible to orient a mobility device so that it is centered on the operable part. Best practice is to locate these operable parts a minimum of 12 inches (30.5 cm) from corners where they will be more easily accessed.

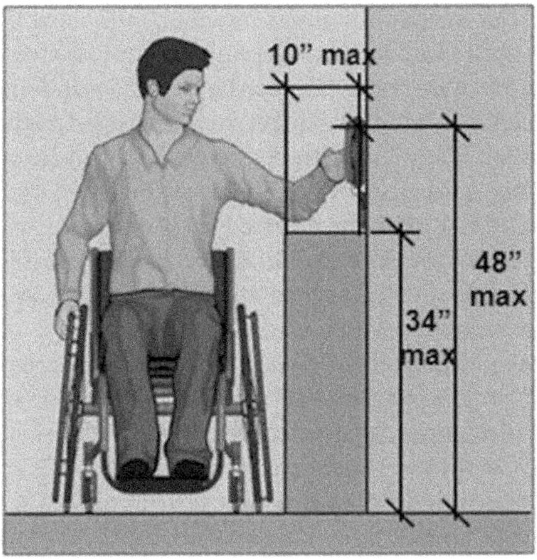

Figure 36. Accessible obstructed high side reach range to an operable part. Reprinted from reference (2).

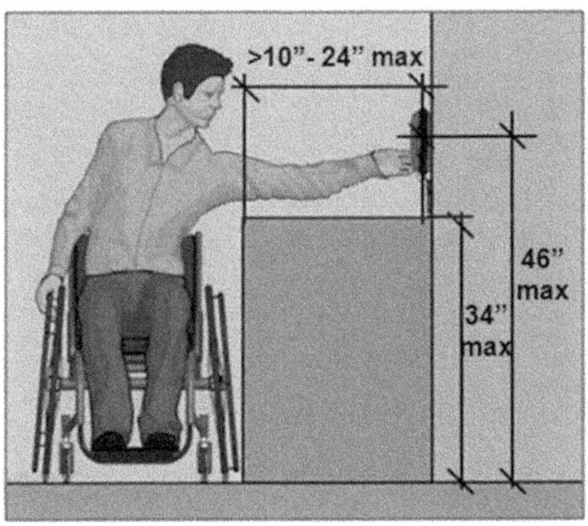

Figure 37. Accessible obstructed high side reach range to an operable part. Reprinted from reference (2).

Limits of Protruding Objects

In order to prevent hazards to people who are blind or have low vision, the ADA Standards limit how far objects may protrude into circulation paths, including circulation paths within lab spaces. These requirements apply to all circulation paths and are not limited to accessible routes. Circulation paths include interior and exterior walks, paths, hallways, courtyards, elevators, platform lifts, ramps, stairways, and landings (*1*).

Many people with visual disabilities travel closely along walls to assist with wayfinding. When objects are mounted on walls, or similar areas such as partitions or columns along circulation paths, they can be hazardous to someone with a visual disability if the leading edge of a projection is mounted higher than 27 inches (68.6 cm) above finish floor and protrudes more than 4 inches (10.2 cm) into the circulation path. This is because something that is mounted higher than 27 inches above the floor would not likely be detected by someone using a cane, given the height of the object above the floor. Additionally, projections into circulation paths below a minimum headroom clearance of 80 inches (2.03 m) above a finished floor can be hazardous to someone with a visual disability.

Objects that have leading edges within the cane sweep area (between finished floor and at least 27 inches (68.6 cm) above finished floor) do not pose hazards to people with visual disabilities, as their cane will detect the elements and they can navigate around these objects. Some common examples of elements that violate this provision include wall sconces that protrude more than 4 inches (10.2 cm), handrails, drinking fountains that are not recessed, cabinetry, and equipment boxes located on walls, such as fire extinguisher cabinets and other types of laboratory equipment. See Figure 38.

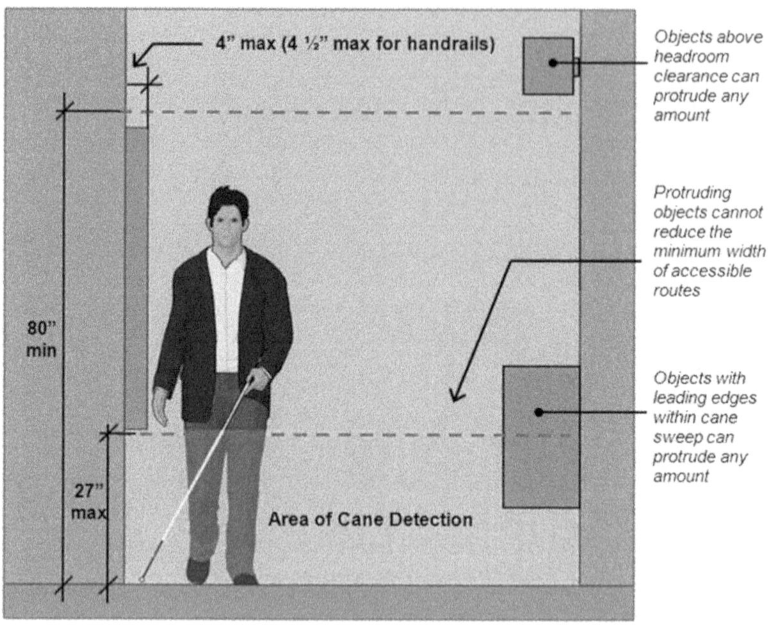

Figure 38. Protruding objects. Reprinted from reference (2).

Figure 39 illustrates an accessible eyewash station/shower and highlights several key accessibility features, including the limits of protruding objects discussed above, as well as accessible operable parts and knee and toe clearance. The eyewash station contains an operable part that is located between 15 inches (38.1 cm) and 48 inches (1.22 m) above a finished floor. This triangle design does not require any tight grasping, pinching or twisting of the wrist to operate and can be used with one hand. Note that clear floor space is provided in front of the unit and the operable part for the eyewash is a "paddle style" design which can be easily operated. There are 27 inches (68.6 cm) of knee clearance underneath the basin, which complies with the limits of protruding objects and also provides accessible knee clearance underneath the unit to permit a person that uses a wheelchair to maneuver.

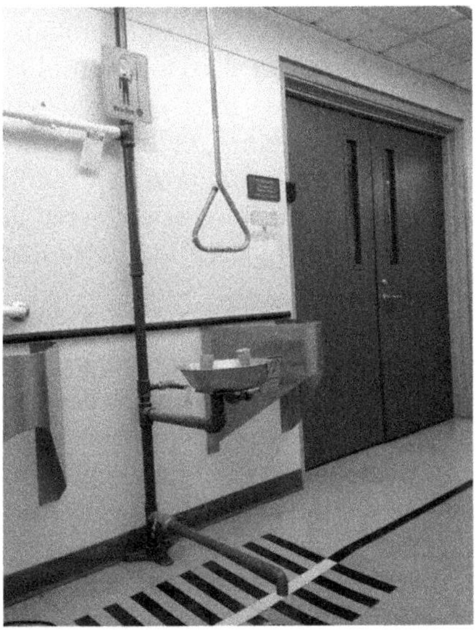

Figure 39. ADA compliant emergency shower and eyewash. Published with permission from Ellen Sweet, editor.

Laboratory Surfaces: Perception of Limits, Planes, and Distances

Concepts of visual perception, similar to those for signage, apply to laboratory enclosures. The goal is to improve overall visual perception within the lab environment. Persons with limited vision can better discern vertical and horizontal planes, edges, distances between planes, and changes in direction by applying contrasting pale and darker colors on surfaces of laboratory walls, floors, bench tops and benches. For example, if a laboratory bench top is dark in color, as several commonly used materials are (e.g. epoxy, slate, and soapstone), then cabinets and supporting elements below should be manufactured in a distinctly paler shade or color that further distinguishes horizontal from vertical planes. To assist persons to better perceive the distance between a countertop and floor, install flooring material with a distinct pattern or with different hues from the cabinets and countertops.

Raised edges are another signal that can be perceived visually or tactilely where countertop surfaces end. Raised edges can be integrally molded or separately applied to the front of the countertop. Bull-nose edges of sealed wood or chemical resistant molded plastic or neoprene can be applied; integral seamless marine edges can be manufactured in stainless steel and other seamless countertop materials. For example, people with limited vision can feel along a smooth ridge-edge, before confidently placing glassware or equipment onto a bench top.

Wall surfaces in laboratories can assist orientation in laboratories, particularly in large, open concept, laboratories that have many aisles and few walls other than their enclosures. Surfaces and frames, surrounding the exteriors of windows, can reduce glare when coated with white or pale-color, non-glossy paint. Glossy coatings on walls increase glare from sunlight through windows even when window coverings, shades, or blinds are used to control light. Pale, matt coatings distribute incoming light from windows more evenly to reduce glare.

This concept applies to laboratory ceilings as well, especially when coordinated with electric light fixtures that provide even illumination using both direct and indirect fixtures. White and very pale-colored matt-finish ceilings can be designed to reflect natural sunlight deep into laboratory spaces. This saves energy and reduces sunlight's glare.

Surface coatings on windowless walls can be designed to improve signage visibility and make signs more noticeable in laboratory environments. Solid pale neutral color, matt coatings work well to increase perception of light by gentle diffusion. Heavily patterned surfaces with paint or other coating can distract workers in the laboratory looking for safety and egress signage mounted from ceilings or walls. Safety and egress signage can be hard to locate in visually "loud" interior environments.

In addition to good egress signage, exit and entry doors with windows and metal or wood coated with bright colors can help capture lab occupants' attention and draw their eyes to where they can safely exit. This tactic also applies to fire extinguishers that are traditionally painted fire engine red. Bright, intense yellow can be applied in areas to reinforce attention to official caution signs and other safety features or equipment (*5*). Examples of this include using circles of bright yellow beneath emergency showers and eyewash locations to bring workers' attention to where to find this equipment (*8*).

Electric Lighting

Lighting fixtures and their layout are important factors in laboratory design for the productivity and safety of all workers, but should be addressed specifically for persons who have vision disabilities or who have hidden disabilities where stimuli may be an issue. In order to provide adequate levels of light at laboratory bench tops and equipment, engineers use building codes, lighting industry standards, and technical manuals to guide them in selections, specifications, and distribution of lighting fixtures installed in ceilings and on walls. A lighting design that addresses accessibility of those with vision or hidden disabilities should take special measures to ensure that once the occupants are using the space, they do not experience inadequate lighting of the space following construction. Two strategies can mitigate problems in lighting design.

The first consideration is to contract with the lighting designer to provide a complete lighting simulation on 3-D computer programs, including a table of calculated light levels in all portions of a typical laboratory. One alternative method is to require, in the contract with the constructor, that the contractor builds a full-sized laboratory mock-up that includes the engineers' specified fixtures and the distribution as drawn in the construction documents. Without all the equipment, apparatus, and supplies that will eventually fill the laboratory bench tops, open shelving above and on walls, this lighting simulation will be just that, a simulation.

A mock-up can be furnished with volumetric simulations of walls, wall shelving/cabinets, typical casework, large equipment, and storage units as shown in the laboratory design documents. Large volumes can cast dark shadows that decrease visual acuity, if lighting is uneven. With walls and volumes defining the space within a laboratory, lab occupants and facilities managers can walk through the proposed space to more accurately evaluate the adequacy and effectiveness of the lighting.

The second consideration is to let the laboratory users locate task light fixtures to meet their specific needs. The owner or users can provide flexible and/or moveable task lighting fixtures to work areas. Moveable-arm drafting lights have shades that reduce glare and focus light on circumscribed areas. The types of light bulbs, magnifying lenses, or light strips that meet the needs of occupants with vision impairments can be provided. Support clamps for moveable-arm task lighting can be installed on most shelving systems or supported by clamps and bases that can be moved to the most helpful position along the bench top or on equipment.

Movable lighting fixtures are considered "furniture and fittings" and are normally not included in the construction contract. However, the power outlets should be on engineering contract documents and part of the construction contract. The owner or users often purchase these out of the owner's contingency funds for furniture.

An alternative to flexible/moveable task lighting is to install fixed under-shelf/cabinet lighting. These fixtures have on-off switches at the bench, separate from overall lab wall-mounted light switches. Construction contractors can install them as part of the cost of the project if they are shown in the drawings and specified by the lighting designer. Unfortunately, under-shelf light can be obstructed by small equipment and devices placed on the bench top; this defeats its purpose. Under-shelf fixtures do not adapt well to specific needs of laboratory workers with vision limitations.

In laboratories and laboratory buildings special light fixtures are used specifically for illuminating life-safety signs and other safety communication devices. Other signs often have internal light sources, such as most EXIT signs, and caution signs, such as "Laser-in-Use" signs. Safety signs and illumination of signs must meet OSHA, local and national building codes, and industry lighting standards. This applies to signs used for life safety: exits, fire stair locations, required egress pathways, and areas for fire-refuge (*8*).

Life safety challenges that occur in laboratory buildings and cause stress or disorientation of people are:

- visibility obscured to way-finding to lab exit doors by smoke or release of irritating chemicals;
- black-out conditions during power failures and;
- emergencies that require immediate evacuation;
- Unplanned events that obscure vision.

When smoke or irritating fumes fill the air in a lab, occupants should drop to the floor to avoid smoke and fumes. Directional light fixtures should be installed on laboratory casework and walls that assist occupants with egress. Special light fixtures mounted low on walls leading to exit doors are also good cues for people to realize a door is near and to differentiate them from lighting along typical laboratory work aisles.

Another approach to mitigate risk of blackout conditions in large laboratories is to supply emergency power to selected light fixtures on the ceiling. These fixtures can minimally illuminate the lab so workers can safely shut down hazardous equipment or processes, then to leave the lab. Emergency power to selected general light fixtures is legally required in some building codes to maintain minimal light along corridors, stairs, and other egress pathways in buildings.

Building codes, which are state specific, usually do not require emergency general lighting or special egress-path lighting within large laboratories. The laboratory risk evaluation guides the owner who makes that choice to improve safety and pay extra for additional lighting. Dim light may not be adequate for workers with limited vision to evacuate a laboratory safely alone, but others in the lab may be able to see and can assist workers who need help exiting. Safety signs on egress pathways (EXIT signs, etc.) must be continuously illuminated by battery and/or emergency power, as regulated by most local, state, and national building codes.

Fire alarm systems and equipment are required to include visual (flashing lights) and audible (siren or horn) signals to alert laboratory occupants with sensory disabilities. In addition, public announcement systems can be specified for auditory as well as digital sign to display messages to assist with safe egress.

Laboratory Chemical Hood Safety and Testing

An ASHRAE 110 test confirms the "As Installed" conditions following installation. Annual performance inspections of laboratory chemical hoods represent "As Used" conditions. But, these are not as comprehensive as ASHRAE 110. Conditions of hood use that effect containment often surround the clutter of equipment, gas cylinders, and chemical bottles that can obstruct airflow into part of the hood's sash opening where they can disrupt airflow into and within the hood chamber. These conditions can block airflow and result in leakage of airborne contaminants which cause unsafe conditions for users who are seated at the hood (*9*).

Chemical hood users who work in a seated position or who are short in stature, are at greater risk for chemical exposures from hood leakage, than workers who stand and are 5 ft. 10 in. tall (1.78 m) or taller. A National Institutes of Health (NIH) study modelled 250 laboratory and ventilation configurations for chemical hoods that used computational fluid dynamics (CFD) techniques (*10*). These simulation results were empirically confirmed in on-site conditions. Both methods quantify the leakage of contaminants that can widely disperse in a laboratory under specific facility conditions. The breathing zone of an operator in a seated-position is within the range where the NIH study quantified leakage and escape of measurable volumes of contaminated air that is generated within a hood chamber. Leakage can occur when a person works in the hood chamber and is due to movement of the operator and objects being manipulated and stored inside. Escape from the hood chamber also occurs when a person walks past creating a cross-draft. This movement causes air turbulence that may not be recaptured and drawn back into the hood. This NIH study offers extensive quantitative data to fully understand risk factors for those persons with physical limitations who must sit while conducting chemical experiments in chemical hoods. The open portion of a chemical hood sash is too close to the breathing zone of the seated hood-operator. This may be mitigated by installation of a combination sash that the user can place between them and what they are working with on the bench. But, a modified ASHRAE 110 test would need to be conducted to determine containment in As-Used conditions.

Conclusion

Laboratory planners, engineers, and architects will have to modify layouts for new science facilities to work seamlessly for persons entering science careers with disabilities and scientists who experience emerging functional limitations as they age. Old rules-of-thumb may offer inadequate responses to new workforce demographics. Attention should be taken to select as many laboratory components as possible that will enhance sensory and physical performance and safety for disabled workers.

References

1. 2010 ADA Standards for Accessible Design, U.S. Dept. of Justice (DOJ). https://www.ada.gov/2010ADAstandards_index.htm.
2. "Guide to the ADA Standards". https://www.access-board.gov/guidelines-and-standards/buildings-and-sites/about-the-ada-standards/guide-to-the-ada-standards.
3. International Codes Council, ANSI A117.1 Standard-2017 Accessible and Usable Buildings and Facilities. https://codes.iccsafe.org/public/document/ICCA117_12017/foreword (accessed November 13, 2017).
4. United States Department of Labor Occupational Health and Safety Administration, General Industry Standard 1910, Subpart E- Means of Egress. https://www.osha.gov/pls/oshaweb/owasrch.search_form?p_doc_type=STANDARDS&p_toc_level=1&p_keyvalue=1910 (accessed November 9, 2017).

5. United States Department of Labor Occupational Health and Safety Administration, General Industry Standard 1910, Subpart J-General Environmental Controls. https://www.osha.gov/pls/oshaweb/owasrch.search_form?p_doc_type=STANDARDS&p_toc_level=1&p_keyvalue=1910 (accessed November 9, 2017).
6. DiBerardinis L.; Baum, J; First, M.; Gatwood, G.; Seth, A. *Guidelines for Laboratory Design: Health, Safety, and Environmental Considerations*, 4th ed.; Wiley: Hoboken, 2013.
7. New England ADA Center, ADA Checklist for Existing Buildings, Priority 2: Access to Goods and Services. https://www.adachecklist.org/checklist.html (accessed December 21, 2017).
8. United States Department of Labor Occupational Health and Safety Administration, General Industry Standard 1910, Subpart G- Occupational Health and Environmental Control. https://www.osha.gov/pls/oshaweb/owasrch.search_form?p_doc_type=STANDARDS&p_toc_level=1&p_keyvalue=1910 (accessed November 9, 2017).
9. American Society of Heating, Refrigeration and Air-Conditioning Engineers. *ANSI/ASHRAE Standard 110-2016 Methods of Testing Performance of Laboratory Fume Hoods*; ASHRAE: Atlanta, GA, 2016.
10. Memarzadeh, F. *Methodology for Optimization and Testing of Laboratory Hood Containment*; NIH: Bethesda, MD, 1996; Vols. 1 and 2.

Chapter 4

Accommodating Persons with Physical Disability in the Lab

LaWanda H. Cook*

K. Lisa Yang and Hock E. Tan Institute on Employment and Disability, Northeast ADA Center, Cornell University, 311 Dolgen Hall, Ithaca, New York 14853-3901, United States
***E-mail: lhc62@cornell.edu**

Physical impairments can impede movement, coordination, or sensation, and result in weakness and lack of muscle control. These limitations can affect the upper and/or lower body and can result from a variety of congenital and acquired disabilities. These disabilities may be readily apparent, such as when an individual uses crutches or a wheelchair for mobility. Other conditions may be hidden or invisible, for example when someone has limited hand strength due to arthritis. In this chapter we consider modifications to the lab that may benefit individuals with a range of physical limitations.

Prevalence of Physical Disability

Physical disabilities can limit a person's functioning in many ways, and individuals may have difficulty walking, caring for personal needs, or living independently. Based on analysis of the 2015 American Community Survey (ACS) Public Use Microdata Sample, there are approximately 13.5 million working age adults (18-64) in the U.S. who report having disabilities that limit their functioning in one or more of these areas (*1*). Although such limitations can be caused by different types of conditions, many of these individuals have physical disabilities.

Some physical limitations are due to a condition present at birth, such as cerebral palsy, spina bifida, and dwarfism. Others, like multiple sclerosis (MS), may manifest later in life. Still, others may be the result of injury or illness, such as spinal cord injury, limb amputation due to diabetes or cancer, or limited use of an arm or leg as a consequence of a stroke. Whatever the cause of one's impairments, there are a number of ways to accommodate and provide an inclusive educational experience in the lab for students with physical differences.

Non-Discrimination and Equal Opportunity

Under the ADA, no qualified individual with a disability shall, on the basis of disability, be excluded from participation in or be denied the benefits of the services, programs, or activities offered, or be subjected to discrimination by any public or private entity. Eligibility criteria and other practices must not discriminate against a person solely because of disability. An example of a potentially discriminatory policy would be requiring all students who purchase books at a school store by check to show a driver's license, because some disabilities preclude a person from driving. An alternative policy would be to require students to show a State issued identification (this would include a driver's license or any other State-issued photo ID, such as a DMV issued identification card for people who do not drive).

Program requirements, including university eligibility criteria and course requirements, should not include specific capabilities that are non-essential or that can be modified. For instance, a requirement that students have the ability to carry 10 pounds may be discriminatory, assuming that carrying 10 pounds is a non-essential task that rarely needs to be completed. It is reasonable that this task could easily be accommodated by assistance from another student. Similarly, if it was an essential task, and if the items could be moved using a cart or other equipment, rather than being carried, the student could be provided with a cart as an accommodation.

All students – with and without disabilities – need to be aware of risks associated with the hazards present when working in the lab and how to engage in safe practices. This is as much a part of their education as learning the course material. In situations where a person with a disability is perceived to themselves be a hazard and present a significant risk of substantial harm to the health or safety of self or others, an entity is required to first conduct an individualized assessment prior to a stated decision to deny or discontinue services. This is similar to a risk assessment that would be conducted prior to any laboratory experiment involving chemicals, but with the additional expertise of someone familiar with determining disability related accommodations. The individualized assessment must consider current medical knowledge or the best available objective evidence. This evidence is used to determine the nature, duration, and severity of the hazard; the probability that the potential injury will actually occur; and whether reasonable modifications of policies, practices, or procedures or the provision of auxiliary aids or services will mitigate the risk.

If reasonable modifications can minimize or eliminate the risk, these must be tried, as the intent of the ADA is to promote the most integrated participation and inclusion of individuals with disabilities. Under the ADA, students with disabilities are to be provided the opportunity to participate alongside their peers without disabilities, unless an individual is unable to do so, even with reasonable modifications to policies or practices, or the provision of auxiliary aids. If a person has limited mobility of the upper or lower extremities, full participation in an integrated session is possible given that the space is accessible as required by the ADA, and any furniture and equipment is universally designed or modified based on individual needs.

ADA Accessibility Standards

A positive lab experience begins with the ability to access the building where the lab is housed. For individuals with physical limitations, key accessibility features include the availability of a stairless or ramped entrance, doorways that are wide enough to accommodate wheelchair users, automatic door openers, and lever handles that allow for safe, independent entry. Similarly, the entrance to the lab itself, must also take into account the needs of individuals with physical limitations. These features should be consistent with the ADA Standards for Accessible Design, to ensure, for example, that there is sufficient room for a wheelchair user to navigate within the lab (*2*). While building accessibility has been in place for many years, there might still be older academic buildings that are inaccessible. For complete, detailed information about accessibility requirements, which are discussed in Chapter 3 of this text, see the ADA Standards.

Universal Design

In regard to physical access to the laboratory environment, adherence to the principles of Universal Design (UD) allows greater access for everyone. UD refers to "the design of products and environments to be usable by all people, to the greatest extent possible, without the need for adaptation or specialized design" (*3*). One example of a UD feature that can meet the needs of people with a variety of physical limitations is an adjustable height workstation which can accommodate individuals who use wheelchairs, people whose health conditions require that they sit to perform tasks, and individuals of various heights. Individuals with physical limitations of the upper or lower body may also require more time to complete tasks and UD features help to enhance productivity and reduce fatigue.

For individuals with upper body impairments, as well as students who must work from a seated position, needed supplies should be within reach. This includes all required safety equipment, drawer pulls, fire extinguishers, and other work and emergency equipment. Adjustable seating is also helpful. The use of lighter weight materials, such as plastic rather than glass, can reduce the effort needed to complete tasks.

Individual Accommodation under the ADA

Settings that incorporate universal design principles and are in compliance with ADA Standards for Accessible Design will meet the needs of most people (*2*). However, in some instances, individual accommodations may be required.

In such cases, the goal should be to ensure that the student with a disability has the opportunity to learn the same content and have the same experiences as the other students. All students must be able to demonstrate mastery of course material. For students with disabilities, accommodations that involve receiving or demonstrating understanding of information in ways that are different from their nondisabled peers, may be necessary. Accommodating people with disabilities requires an interactive discussion that leads to an understanding of the impact of the disability on the activities required for their school program and/or employment.

The educational institution should have a clear process by which accommodation requests can be made. This process should be well-documented and widely communicated to all students, staff, and faculty. Such a process should be in line with the ADA and should clarify the rights and responsibilities of both the individual and the institution.

Under the ADA, a student requesting an accommodation is responsible for disclosing that they have a disability which impacts their participation in their educational program. While the decision to disclose and the timing of disclosure are up to the individual, students should be aware that institutions are generally better able to meet their needs with sufficient advance notice. Many conditions that impact strength and functioning of the upper or lower extremities are not readily apparent. In cases where the disability is not evident and/or the need for accommodation is not clear, the institution may request medical documentation in order to confirm the condition and determine a reasonable accommodation. In many cases, disclosure and the confirmation of disability is handled by a department dedicated to ensuring access for students with disabilities, on most campus this resource is the Office of Students with Disabilities (OSD). Depending on the hazardous materials that will be used and what the accommodation entails, university or departmental safety experts can be helpful in such discussions.

When determining what accommodation is "reasonable" an institution should give consideration to the ways in which the person has been supported or accommodated in completing similar tasks in past classes, or in other contexts. The institution must provide an accommodation that *effectively* addresses the challenges presented by the individual's disability. Because a person's situation can change over time, it is important that the institution periodically check with the individual to see if the accommodation provided still effectively meets the student's needs. In the laboratory setting the re-assessment of risk is done when there is a change to the procedure. There are many types of changes, such as the scale-up in volume of chemicals used and changes in chemicals with slightly different chemical structures that carry different hazards; and students that have varying knowledge of chemical hazards and risk mitigation. Periodically reassessing the suitability of an accommodation that has been provided for a student is part of supporting the student's success. In the laboratory setting, this should be done more frequently, as each lab module and course is different.

The OSD can help to verify needs and offer suggestions for accommodation. For an example of the types of issues to consider when documenting a students' accommodation needs, see the University of California Davis' Department of Chemistry's guidance on accommodating students with disabilities in chemistry teaching labs at: https://chemistry.ucdavis.edu/sites/g/files/dgvnsk196/files/files/page/needs-assessment.pdf.

Accommodating Student Employees

These same principles apply when a student is employed in a laboratory setting and requires a reasonable accommodation to perform essential job functions. In such instances, the institution must engage the student employee in an interactive process to gather needed documentation of the disability and explore accommodation options. As the employer, the institution would ultimately determine an accommodation that will effectively aid the individual in meeting the essential functions of the job.

Accommodating Specific Physical Disabilities in the Lab

Each student is unique. Even students with the same disability can have very different needs. Below are some suggestions for accommodating specific physical limitations that a student may experience.

Difficulty Carrying Items while Using a Wheelchair

Non-slip boards can be placed on one's lap and used to hold items such as tools, liquids, and books.

Difficulty Gripping

Plastic grips with a grooved inner surface can be mounted onto equipment for non-slip grip. Velcro can be helpful to reduce accidental slipping or falling of certain glassware. Plug pullers attach to various types of plugs and have two large finger holes for gripping. Foam tubing placed around laboratory tools enables individuals with limited coordination, strength, and gripping ability to use items with greater control. A Universal Turning Handle can make it easier to turn on taps open/lock doors, and turn knobs on equipment.

Difficulty Reaching

Reachers or grabbers enable individuals to reach and hold onto items. Adjustable workstations make it possible for people of different heights to set up their work area in such a way that needed supplies and equipment are close by. Lowered countertops enable individuals who are shorter or working from a seated position to have items close at hand. Lazy susans keep supplies handy and easily accessible, as they require minimal effort to rotate individual storage compartments toward the user.

Difficulty Standing

Stand/lean or sit/stand stools offer adjustable seating and the option of positioning the stool so that the person can lean or rest against it without being fully seated. Anti-fatigue Mats help to reduce the back, knee, and neck soreness.

The ability of individuals with physical disabilities to complete tasks may also be influenced by factors other than their specific physical challenges. For example, for individuals with conditions such as arthritis, room temperature can increase muscle tightness, which can be both distracting and uncomfortable. Whenever possible, requested adjustments in room temperature should be made as an accommodation. In the event that the work being completed or chemicals being used require a particular temperature setting, the person should be allowed to take other measures (such as using a personal cooling device or wearing a jacket) to minimize the negative impact on functioning.

Another consideration is the amount of **stress** an individual is managing. When working in the laboratory, stress can raise risk when people rush to complete their work or are distracted or tired. These conditions can result in errors or unwanted events such as chemical spills. Stress is a part of life, whether or not one has a disability. Therefore, all students and employees should be aware of available support services. This is particularly important for individuals with physical disabilities, as concerns about school, work, relationships, health, or other issues, can exacerbate the impact of some conditions.

Everyone's definition of and response to stress is different. For many people, habits such as eating a healthy diet and getting sufficient sleep can help minimize feelings of stress. It is important that individuals with disabilities have equal access to all of the support services available on campus. This includes access to physical and mental health services, recreation facilities and programs, and wellness initiatives. Staff should be culturally-competent in disability, as well as other types of diversity, and equipped to serve students with a variety of needs.

Case Studies

As indicated throughout this chapter, many of the needs of laboratory students with physical disabilities can be addressed through thoughtful design of the space. In instances when a specific, individual accommodation is needed, students should be encouraged to work with the OSD or similar campus resource to identify possible accommodations.

Case #1: Burt has a physical disability that limits his fine motor skills. He will be taking a laboratory based course in the spring and has come to you, the professor, to discuss possible accommodations. He is requesting that a personal aide be hired to assist him in the lab, like he had in high school. What do you do?

Case #1 Recommendations: Take some time to find out what specific types of tasks/functional activities are challenging for Burt. Encourage him to work with the OSD to verify his disability and identify possible accommodations. While a lab assistant may be an appropriate accommodation in some instances, there may be other ways to address Burt's needs. Concerns about picking up beakers and other items could be addressed by adding handles to the equipment. Knobs on doors, faucets, and equipment could be replaced or covered with universally designed knobs. And non-slip mats on the countertops would help to keep items in place while he works.

Case #2: Alice has a condition that causes fatigue. She has come to talk with you about how to minimize physical effort while in the lab.

Case #2 Recommendations: Take some time to find out what specific types of tasks/functional activities are challenging for Alice. Encourage her to work with the OSD or similar campus resource to verify her disability and identify possible accommodations. Issues of fatigue might be addressed through the use of an anti-fatigue mat and by moving frequently used equipment within close reach. A sit/lean stool would enable her to work from a seated or standing position; and storing supplies in a Lazy Susan would allow her to get to needed items with minimal effort. Allowing breaks throughout her time in the lab would also be helpful. Scheduling her laboratory at a time of day when her fatigue may be less impactful on performance may also be helpful.

References

1. Erickson, W. Cornell University, Ithaca, NY. Personal communication, 2017.
2. 2010 ADA Standards for Accessible Design, U.S. Dept. of Justice (DOJ). https://www.ada.gov/2010ADAstandards_index.htm (accessed January 15, 2018).
3. North Carolina State University Center for Universal Design. https://www.ncsu.edu/ncsu/design/cud/about_ud/about_ud.htm (accessed October 23, 2017).

Chapter 5

Hidden or Invisible Disabilities and Laboratory Accommodations

Chris Sweet*

Northeast ADA Center, K. Lisa Yang and Hock E. Tan Institute on Employment and Disability, Cornell University School of Industrial and Labor Relations, 312 Dolgen Hall, Ithaca, New York 14853, United States
***E-mail: cs986@cornell.edu**

Hidden or invisible disabilities are those that are not readily apparent to the observer and encompass a wide array of conditions. Like many disabilities they may have been acquired at birth or been obtained at a later time. Since their condition is not apparent to others, people with this type of disability can often feel a greater sense of stigma as their condition is not apparent to others and often feel as though others may think they are faking or dramatizing a condition. Chemistry laboratories are challenging environments and it may not be readily apparent what barriers are present for a student with a hidden disability as they acclimate themselves into a program. In some cases, the student themselves may not understand how their disability is impacting their progress.

In this chapter we will look at a variety of factors and challenges that students with hidden disabilities might face in the laboratory. Suggestions for engaging stakeholders in conversations to meet the accommodation needs of students with hidden disabilities will also be covered.

Introduction

Since the passage of the American's with Disabilities Act (ADA) in 1990 (*1*), there has been enormous progress made towards improving physical accessibility for those with conditions that impair body movement and control. Now, not only do we see accessible ramps at entrances, accessible public bathrooms, and design enhancements that allow for easy reach ranges, and hardware that allows a user to more easily open a door present in new construction, but many older buildings have been upgraded to allow greater access for all.

It is often what we observe that allows us to understand the needs of a person with a disability. Physical space can be altered to meet the individual's needs through the use of universal design standards. Wide accessible parking areas close to entrances and accessible routes of travel allow the user to access buildings in an effective manner and allow for public areas to be open to most. However, when an individual has a disability which is non-obvious, hidden, or invisible, identification of accessibility needs can also be non-obvious to both the observer and the person with the impairment. For these individuals, accommodations are not as clear cut as changing physical features of a workspace.

Once it has been identified that a barrier exists for a student with a non-obvious disability, a dialogue between the student and relevant parties can proceed to identify what the barrier is. At this point, strategies as to how the student can achieve the identified goals of the course can be developed. The planning should include the institutional disability services, often called ODS, the lab professor and/or student teachers, staff from Environmental Health and Safety, and any other stakeholders that have a direct impact on the overall function of the lab. "Disability scholars argue that disability is a social construction, and have summarized it this way: Disability is often less about physical or mental impairments than it is about how society responds to those impairments" (*2*). Educators must respond so as to consider student success and be responsive to all of the needs of the student whether apparent or not.

Examples of Hidden Disabilities

Hidden disabilities can be attributed to many types of disorders. While the list in this chapter is not exhaustive, it will attempt to give a general overview of conditions that are considered hidden or invisible disabilities. While there are similarities that are experienced for each type of disability, the condition will uniquely affect an individual. It is important not to make assumptions that an experience will be the same for every person. Three types of conditions will be discussed here; immune system disorders, brain disease or injury, and genetic disorders.

Immune System Disorders

Allergies

Immune system disorders are commonly seen in the form of allergies during certain times of the year when pollen counts are high. Often medication can mitigate the effects and the sufferer can function without difficulty. Other allergic reactions can be more problematic. In some instances these reactions can have an adverse effect on breathing by triggering asthma attacks in people that suffer from this respiratory condition. Other adverse reactions could be mistook as an exposure to a chemical and cause a response with the emergency medical staff. This could continue occurring causing disruption to the entire class and embarrassment of the student. These can include "absent mindedness, dizziness, trouble swallowing, nausea, constipation, diarrhea, skin rashes, vertigo, vomiting, and other serious cases of health issues that may spring from how the body deals with the allergen itself" (*3*). Allergens in the lab could range from something as simple as latex gloves to the host of chemicals used in teaching. Adding a statement of common allergies into the first day comments or a statement to the lab text book can alert a student if they know they have had a history of problems with that allergen. This way, a student can confidentially speak with the instructor to discuss an accommodation.

Autoimmune Diseases

These diseases result when "the immune system attack(s) the body's own organs, tissues, and cells. Some of the more common autoimmune diseases include type 1 diabetes, rheumatoid arthritis, systemic lupus erythematosus, and inflammatory bowel disease" (*4*). There are a whole host of issues which might affect a student's ability to function well in the laboratory setting that are related to autoimmune disorders ranging from chronic muscle and joint pain, blood disorders, and endocrine disorders.

Brain Disease or Injury

Mental Illness

Mental disorders are a common disability, often hidden, that affect a large portion of the United States population. According to the National Institute of Mental Health, "in 2015, there were an estimated 43.4 million adults aged 18 or older in the United States with AMI (any mental illness) within the past year. This number represented 17.9% of all U.S. adults" (*5*). Mental illness can encompass a wide range of conditions such as depression, anxiety, attention deficit hyperactivity disorder, bipolar disorder, schizophrenia to list a few. While many mental health conditions are managed through therapy or medication, there may be instances where concentration is affected, memory issues can arise, and organization and time management could be impacted. Stress resulting from the illness could also play a role in overall performance.

Traumatic Brain Injury

Traumatic Brain Injury (TBI) can be the result of numerous causes. "According to the CDC (United States Centers for Disease Control and Prevention), there are approximately 1.5 million people in the U.S. who suffer from a traumatic brain injury each year" (*6*), and "in the U.S., more than 5.3 million people live with disabilities caused by TBI" (*6*). When thinking of brain injury causality one can look to the national dialogue around sports injuries caused by blows to the head, yet brain injury can occur from other trauma such as hypoxia (lack of oxygen), chemical exposure, tumors, infection, stroke, and other means (*6*). Effects of brain injury can be various including cognition and memory issues, delayed thought processing, balance issues, behavioral changes, and other non-obvious impacts.

Genetic Disorders

Autism Spectrum Disorder

Autism Spectrum Disorder (ASI) encompasses a range of disabilities which impact the person's social interactions. "Autism is a neuro-developmental disorder characterized by impaired social interaction, verbal and non-verbal communication, and restricted and repetitive behavior. Parents usually notice signs in the first two years of their child's life. The signs typically develop gradually, but some children with autism will reach their developmental milestones at a normal pace and then regress" (*7*). While autism can affect an individual in common social settings the impact in the classroom can be more challenging for the student. "[students] on the autism spectrum may have trouble understanding or communicating their needs to teachers and fellow students. They can have difficulty understanding some classroom directions and instruction, along with subtle vocal and facial cues of teachers" (*8*).

Considerations for Accommodation

Section 504 of the Rehabilitation Act of 1973 (Section 504) and Titles II and III of the Americans with Disabilities Act of 1990 (ADA), prohibit discrimination based on disability in public and private post-secondary institutions (*1*). This is discussed in detail in chapter 2 of this book. In the post-secondary setting the responsibility of overseeing and implementing these accommodations often falls to the Office of Student Disability Services (ODS) or to the Office of Student Affairs. While staff in these offices may be trained in understanding the rights of students with disabilities as well as the obligations of the college under the ADA and Section 504, they cannot be assumed to have an understanding of specific classroom environments and, for the sake of this chapter, the laboratory environment.

University laboratories often have multiple partners that have a collective responsibility of laboratory operations and safety. While the professor may be the lead driver in lab usage, the stakeholders previously mentioned may all have a say concerning the accommodation ideas that are possible and may also contribute to creative ways in meeting the needs of students with disabilities. Effective dialogue between all parties needs to be a part of the process when discussing accommodation ideas. It cannot be assumed that each department understands the obligations under the ADA and Section 504 when it comes to serving a student with a disability.

It may fall on the ODS or to the Office of Student Affairs to educate all parties involved as to the effects of disability types on the learning process as well as to assist the student in beginning a dialogue to formulate what accommodation ideas may be effective for the student without altering the fundamental structure of the class. These offices can help to ensure the students gets the modification they need to be successful while honoring the legitimate concerns of other parties and maintaining student confidentiality in particular as it concerns student medical records relative to the accommodation.

Accommodation Ideas

As previously stated, the above summary is not an exhaustive list of conditions that result in hidden or non-obvious disabilities. The overall thing to consider is that there is a wide range of occurrences that result in disability. Open dialogue when considering the impacts on the learning environment and the learning process by people that have been affected by disability is integral. Accommodating specific needs of individuals will be highly dependent on the challenges that the individual experiences during the learning process. Some scenarios are provided here that show how the learning environment could be altered to meet the needs of the student.

Scenario #1

Julie is a TBI survivor and at times has trouble focusing on verbal directions during lab instruction. She has trouble with focusing on the instructor in front of the class and then shifting her focus back to the experiment they are working on. She is wondering if she could have a teaching assistant work with her at her lab station and go over the steps with her.

Recommendations: It is important to discuss with Julie and other parties what specific issues may be contributing to Julie's inability to shift focus between class instruction and her specific lab work. A potentially effective solution could be alterations to her work space or the way instruction is provided. One possible solution could be to move her lab station closer to the professor. Another option may be to have specific instructions outlined in written format that could help her focus on lab tasks. Future assignments on the syllabus should be evaluated to mitigate any issues that might arise in upcoming laboratory work. However, after completing the interactive process, it may be determined that her original request for a teaching assistant is the best accommodation. However, there are regulatory restrictions on the number of people that can be in a laboratory. If the section is at capacity, an assistant cannot be added.

Scenario #2

Alex is a student with autism who is finding that the laboratory setting is difficult for him and is causing him to experience sensory overload. This is resulting in him having trouble focusing on the tasks while conducting the laboratory experiment. He asks that he be allowed to do lab assignments when class is not in session so that he has the ability to better focus without the presence of external stimuli.

Recommendations: In this scenario it is important to recognize that the accommodation request has implications for the safety of the student and the due diligence of the university to adhere to best practices for laboratory work. Working alone in a laboratory is discouraged, in general, and so there would have to be a teaching assistant (TA) or instructor assigned for that time period. This also would mean that there would have to be funding to pay the assistant. This accommodation would also mean that there was an available time in the physical space where other lab sections, cleaning, or maintenance was not occurring.

This scenario is an instance where it may require further consultation other than the student, student affairs and the professor. In looking at other options one could look to see if the lighting could be altered, noise cancelling headphones could be provided, a change of seating may help, or the student could be provided intermittent breaks during the lab work to regulate the sensory input. It could be that the original accommodation is achievable and the request granted.

Scenario #3

Susan has a nickel allergy. Previously, she has noticed the response only occurs when she actually comes into contact the metal. At the beginning of the semester she sees one of their experiments will be the synthesis of a nickel complex and she is worried that compounds containing the metal will illicit the same allergic response.

Recommendations: This is a case where the correct personal protective equipment can prevent contact with the compounds in use. If there are still concerns, a TA could be utilized to assist her with the steps where the nickel has to be handled. It is always best to have the student experience as much of the experimentation as possible, but if the allergy is too severe or could be triggered by inhalation, then a "dry lab" might be the solution. In this accommodation, the student is given data or views a video and writes up a report based on this information.

Conclusion

While this is a brief look at accommodation considerations it can show how to approach a scenario when a student has a need. It is important to develop dialogue to get a broader understanding of the barriers for a student. It is also important to involve all parties in addressing those barriers, including the safety professionals. With any request for a modification in the classroom, it is critical to understand the functional limitation the student is experiencing and the accommodation(s) that the environment will allow based on safety and regulation. The ultimate result should be to assist the student in achieving their academic goals.

References

1. Information and Technical Assistance on the Americans with Disabilities Act. https://www.ada.gov/ada_intro.htm (accessed January 3, 2018).
2. Burch, S.; Sutherlans, I. Who's Not Yet Here? American Disability History. *Radic. His. Rev.* **2006** (94), 127–47.
3. Allergy and Allergies: Types and Information. Disabled World. https://www.disabled-world.com/health/respiratory/allergies/ (accessed October 4, 2017).
4. Auto Immune Diseases. National Institute of Allergy and Infectious Diseases. https://www.niaid.nih.gov/diseases-conditions/autoimmune-diseases (accessed October 4, 2017).
5. Any Mental Illness Among U.S. Adults. National Institute of Mental Health. https://www.nimh.nih.gov/health/statistics/prevalence/any-mental-illness-ami-among-us-adults.shtml (accessed October 4, 2017).
6. What are the Causes of TBI. Traumatic Brain Injury.com. http://www.traumaticbraininjury.com/understanding-tbi/what-are-the-causes-of-tbi/ (accessed October 9, 2017).
7. Autism: Symptoms, Facts, and Treatment Research. Disabled World. https://www.disabled-world.com/health/neurology/autism/ (accessed October 9, 2017).
8. Education & Autism Spectrum Disorders. Synapse. http://www.autism-help.org/autism-education-school-effects.htm (accessed October 9, 2017).

Chapter 6

Best Practices for Accommodating Hearing and Visual Disabilities in the Laboratory

Joe Zesski*,[1] and Henry Wedler[2]

[1]Northeast ADA Center, Cornell University, 201 Dolgen Hall, Ithaca, New York 14853-3901, United States
[2]Co-founder, Senspoint Design, 805 West Street, Petaluma, California 94952, United States
***E-mail: jzesski@rilnj.org**

Students with hearing or vision related disabilities face unique challenges in the post-secondary laboratory environment. This can occur from a lack of awareness on behalf of the student, the professor, and/or the office for students with disabilities about how to approach learning in this environment, what kind of accommodations might be possible, or preconceptions that limit a student's capacity to be engaged. These types of barriers to full inclusion can inhibit a student gaining and demonstrating knowledge, and fully participating in lab activity [Burgstahler, S. http://www.washington.edu/doit/making-science-labs-accessible-students-disabilities (accessed November 27, 2017)]. Students with sensory disabilities are often discouraged from studying fields like chemistry because society believes those fields rely so heavily on all of our senses and motor skills being fully intact. In short, barriers which discourage students with sensory disabilities from working in the laboratory environment can be reduced with a better knowledge of how these students may thrive in the laboratory. Key to lessening these barriers is an understanding of hearing and visual disabilities, the variety and range of impact of these general categories of disability, and examining what considerations can help to determine appropriate effective accommodations for the laboratory. In this chapter, we shall explore what is generally meant by a hearing or visual disability and how even a single subset of one of those categories contains a wide spectrum of impact.

When considering accommodations for the post-secondary environment, each individual's disability, the impact of that disability on performing laboratory work successfully, his or her skills, and a given situation's learning objectives need to be taken into account. Faculty and colleagues must hold students with disabilities to the same level of expectation as their peers while recognizing that the techniques or equipment those students employ to accomplish a project, task, or assignment may need to be adapted. "Students with disabilities can work hard and be productive and should be expected to do so" [Burgstahler, S. http://www.washington.edu/doit/winning-equation-access-attitude-success-math-and-science (accessed November 27, 2017)]. We shall then review some possible academic adjustments/accommodations that could assist individuals with hearing and visual disabilities in a laboratory. Following this general discussion, we shall focus on the practical experience of one of this chapter's co-authors, Henry Wedler, as both a student and as a professional in the field. He will share strategies for accommodating students who are blind or visually impaired (BVI) in a chemistry lab.

Hearing and Vision Disabilities; Potential Accommodations

Approximately 2.3% of Americans (7,297,100) have a visual disability and about 3.5% (11,254,700) have a hearing related disability (*1*). Within those populations, there is great variance on what a visual or hearing disability means. And for the purposes of this book, it is indispensable to recall that every student's experience of disability will be unique to that individual. To create some general parameters however, we will examine broad categories within each of these larger classifications of disability.

About Visual Disabilities

Visual disabilities can run the gamut from the complete absence of vision (blindness) to a degree of usable vision (often termed low vision). This is not to say that someone who uses ordinary eye glasses should be considered as low vision. The Americans with Disabilities Act (ADA) states that the ameliorative effects of ordinary eye glasses- the amount of compensation for vision loss that ordinary eye glasses or contacts provide- should be taken into account when determining if someone has a disability (*2*). This is an exemption to the ADA's principle that the effects of ameliorative measures should not be considered in weighing if an individual has a disability. And while the ADA does not specify a definition for low vision but rather gives a broad definition (see chapter 2), there is a commonly held legal understanding (*3*) of low vision to mean a medically diagnosed central visual acuity of 20/200 or less in the better eye with the best possible correction, and/or a visual field of 20 degrees or less. Errors of refraction in the eye, diseases of the eye, and other vision-related conditions often underlie low vision. Vision loss may be genetic or acquired through illness or injury. Possible effects of low vision are not limited to but can include: myopia (near sightedness), hyperopia (far sightedness), loss of peripheral vision (due to Glaucoma for example), cataracts resulting in clouded vision, macular degeneration resulting in a loss of central vision, night blindness, color vision deficiency, lack of depth perception, or floaters (small specks or clouds moving in the field of vision) (*4*). Other effects are possible as well and even two individuals with the same general condition experience their vision differently.

While individuals who are blind or visually impaired (BVI) may share some common strategies for adapting to their environment, not all will have the same comfort or skill deploying certain tools. For example with two students who are blind, one may be comfortable with and prefer to absorb written information through braille while the other may prefer to do so through the use of feedback provided by a screen reader. Yet even with these variations, there are some general accommodation ideas and best practices that can be useful in a laboratory or class setting. First, written information presented in the front of a class or lab should be verbalized. This might be information on a blackboard or screen. When displayed written material is used, a student who has low vision may use a magnifier or binocular to see the print. He or she may need to be given preferential seating near the front of the class or lab in order for her to more readily see visually displayed materials. Written material can also be provided to students prior to the class in braille, large and/or high contrast print, or accessible electronic format. When using technology or scientific equipment, a student who is BVI may use tactile markings (such as braille or another tactilely discernable material), large print markings, screen magnifiers, screen readers or speech enabled equipment, a computer or video monitor to enlarge a microscope images or other equipment output, electronic braille displays, or verbal descriptions of demonstrations or visual aids (*5*). Some students may find raised line representations or other forms of tactile modeling beneficial. As will be discussed later in this chapter, creative approaches on how to convert visually represented concepts into meaningful tactile or verbal content requires creativity and working with the individual to understand how he best learns.

In addition to the above accommodations or academic adjustments, faculty should be aware that students who are BVI may use either a white cane or a service animal in the classroom or in the laboratory. This does not mean that the individual poses a threat to the safety of himself or others. The student has used such means to assist their mobility on a daily basis and is cognizant of their environment. If a blind student requires sighted guidance to negotiate the laboratory, he should make a request for such assistance. A guide dog being used as a service animal is trained to behave appropriately and to lie quietly when not assisting their handler. A student using a service animal has the right to bring a service animal where ever the public is permitted to go. This will also be touched on later in this chapter as well as in a subsequent chapter of this book.

Finally, one more example of a modification for a student who is BVI could be to have a lab partner with whom to work on a regular basis. This kind of collaboration is most effective when it is done with team work, forethought in preparation, and strong communication between the individual with a vision disability and their colleague. Later in this chapter, Henry Wedler will explain his best practices that worked well for him when working with lab assistants. It should be noted here that Henry Wedler is completely blind

About Hearing Disabilities

Having considered vision related disabilities, we will now address students who are deaf or hard of hearing (D/HH). It is important to recall that a hearing related disability can represent a wide range of auditory perception. Individuals may be born with their disability or have acquired it. It can be caused by any variety of factors such as heredity, childhood illnesses, pregnancy-related illnesses, injury, excessive or prolonged exposure to noise, or age. Each individual will have incorporated their disability differently and developed personal strategies for communication. Each individual will experience sound in their own way. "There are variations in how a person becomes deaf or hard of hearing, level of hearing, age of onset, educational background, communication methods, and cultural identity (*6*). How people "label" or identify themselves is personal and may reflect identification with the deaf and hard of hearing community, the degree to which they can hear, or the relative age of onset." For example, an individual who loses their hearing as an older adult may not use or have trouble learning sign language or other form of communication. Conversely, an individual born deaf or hard of hearing may use sign language as their primary mode of communication.

In addition to deaf or hard of hearing being a descriptive term for a range of disabilities, Deaf has a separate meaning that describes a particular culture, experience, and perspective. The term deaf refers to the physical audiological condition while Deaf refers to something larger; a group "who share a language – American Sign Language (ASL) – and a culture (*7*). The members of this group have inherited their sign language, use it as a primary means of communication among themselves, and hold a set of beliefs about themselves and their connection to the larger society." While we shall not explore it in greater detail here as it will not directly relate to providing an accommodation in the post-secondary environment, however the cultural awareness is essential to forming a successful collegial relationship. Faculty must recognize these distinctions to approach the individual respectfully. This can be important in recognizing how an individual understands their situation. For example, we shall briefly describe three broad categories of hearing loss. A Deaf individual may not find such a term truly appropriate to herself as she may not regard her hearing as "lost", but simply part of her identity.

There are four general types of hearing loss; conductive, sensorineural, and mixed. Conductive hearing loss is associated with the outer and middle ear; the bones of the ear, the eardrum, or the membranes that relay sound to the inner ear. Sound is therefore not conducted or conducted fully to the inner ear. Sensorineural hearing loss occurs when the nerves of the inner ear or the hair cells are damaged. Mixed hearing loss describes when both conductive and sensorineural hearing loss are present. Central hearing loss is the result of damage or impairment to the nerves or nuclei of the central nervous system. Each of these has different implication on the type and frequency of sound that an individual is able to perceive and how that person may choose to adapt. For example, a student with a conductive hearing loss may choose to use a hearing aid or cochlear implant. However, another individual with central hearing loss would not benefit from such assistance (*8*). "Hearing loss is generally described as slight, mild, moderate, severe, or profound, depending upon how well a person can hear the intensities or frequencies most strongly associated with speech." With the understanding that a student who is D/HH will have unique needs based on a variety of factors, we will consider some potential accommodations for the laboratory.

Conveying different types of information call for methods appropriate to the circumstance. For example, if a student has difficulty understanding a professor's spoken presentation or direction, several approaches might be adopted. Printed materials or instructions could be provided prior to the class. Communication Access Real-time Translation (CART) is a service that produces real-time captioning of spoken words via a computer screen and could be incorporated where lectures are being used. If a student reads lips as part of their communication method, then the professor should position themselves so that the student can do so while they are speaking. In general however, this is a best practice for accommodating a student who is D/HH who does not read lips as other information, for example facial expression, can convey meaning. The student might also need to be given preferential seating near the front of the room or lab to facilitate this process. Use of an assistive listening system (ALS) such as a FM, infrared, or induction loop system may serve some students. Another alternative may be to provide a qualified sign language interpreter if a student who is D/HH communicates by sign language. If this accommodation is provided, it is important that the interpreter know how to communicate the technical language and terms used in the given course or lab. It may be advisable for the student and interpreter to speak prior to the class or lab to insure there is no confusion. It can be beneficial to provide an interpreter with what information will be covered on a given day in advance. The professor leading the class or lab should also meet with the interpreter to agree on how they will share the physical space in the given room. Also, faculty should be mindful of the need to give verbal space between moments of lecture and a demonstration (*9*). If the professor does not pause his or her speech when conducting a demonstration, then the student relying on the interpreter cannot focus on both the interpreter and the demonstration being shown.

Besides the communication between faculty and student, peer to peer communication is also important to address in the class and lab. Depending on the type and extent of communication needed, different strategies and methods might be helpful. If small group or partner work needs to be done, allowing for best positioning and viewing for the student who is D/HH and colleagues is important. Even if a student who is D/HH does not use lip reading, a great deal of expressive communication is also conveyed by body language. Another helpful technique in a group is to adhere to turn taking with the speaker first raising his or her hand (*10*). This will allow the student who is D/HH to more easily recognize who is communicating what; in addition to improving the overall communication dynamic. Competing voices make communication difficult. Transcripts or notes should be shared following a meeting, discussion, or class. For one-on-one interaction, texting, email, or writing with pen and paper may meet the situation's need. A shared computer could also serve a similar purpose. There are also augmentative and alternative communication devices (AAC) that can assist those students who may have a related speech disability who may need to interact with someone who does not know sign language.

When information is conveyed by video or other multimedia, whether in the classroom or by an online platform, spoken content should be accompanied by captioning. If it is delivered in the class or lab, preferential seating may again be important to afford the student a clear view of the text. If the media appears online, then a transcription in conjunction with captioning may not only benefit a student who is D/HH, but other students that may wish for a supplemental means to absorb information.

In the laboratory, communicating safety information in multiple ways is necessary. Visual warning signals must be integrated into the lab. While some students who are hard of hearing may be able to detect an audible alarm, others may not. Moreover, students who are deaf would not be alerted. In a noisy chaotic environment, having both auditory and visual warnings benefit all. Similarly, safety information should be posted conspicuously thereby making this critical knowledge easily accessible to all including the student who is D/HH. A safety tour for those new to a particular lab can provide another means of sharing this same information. In regard to safety, it is important that an interpreter (if present) also be familiarized with this information and related procedures. Design of the lab space is another consideration. In general, eye clutter in the lab should be reduced as much as possible to allow for a clear line of sight (*10*). Transparent fume hoods are useful for a student who is D/HH or researcher to see more of the lab environment and what is happening within it. Work stations can be arranged to provide as much viewing exposure to the lab as possible.

Remembering General Guidelines

While we have discussed students who are blind, low vision, deaf, or hard of hearing in general terms, keeping each student's unique perspective in mind is imperative. "Barriers to working with individuals who are D/HH in the academic or professional laboratory are typically 'attitudinal' (or founded in general naiveté. related to working with individuals with disabilities), and not usually related to safety" (*10*). The accommodation suggestions mentioned here are intended to give a cross section of possibilities. Students who are D/HH will already have strategies and techniques that they have used and adapted to succeed. Often, a simple question from a professor asking for how the individual would prefer to handle communication or other aspect related to their disability will clear up confusion. The student is much less likely to be nervous in addressing a need than the faculty. A student must still be encouraged to seek accommodations through the student disability services office or similar campus entity. As part of that process, the office will take into account the learning environment and demands, what tasks must be accomplished, what skills are needed to perform those tasks, what aspects or components may require accommodation, and what appropriate academic adjustments may exist (*11*). The office should consult with the student and may speak with the relevant faculty member to create an informed understanding of what is appropriate and necessary.

The experience and research of this chapter's co-author, Henry Wedler, provides insight. As an individual who is blind and who has a Ph.D. in chemistry, he has had to negotiate some of the issues that we have reviewed. He will share what strategies have been successful for him and for those students with a visual related disability who he has mentored in their interest in chemistry.

The Importance of Safety and Practicality in the Laboratory: Recommendations for Training Students with Disabilities in Laboratory Sciences with a Focus on Blind or Visually Impaired Students in Chemistry (*12*)

Background

I was born completely blind. In order to study chemistry productively as a blind student, I have had to develop my own adaptations to maximize my experiences both in the classroom and the laboratory and to maximize my efficiency as a chemist. This background section will focus on my experiences, provide a status update on the ability for students with disabilities to study the sciences independently, and provide a brief summary of best practices that work well for me when working with any assistant. Note that this information is coming purely from my experience so few references will be used to back this preliminary section. Also, since I am blind, we will focus on students who are blind or visually impaired (BVI) studying laboratory sciences.

I became excited about chemistry as a junior in high school in an honors chemistry course. In that course, the instructor and I realized that chemistry was quite visual and potentially impractical for a blind person to study rigorously. I believed that chemistry might be impractical for me but I never thought of it as a visual science. As I pointed out to the instructor near the beginning of the second semester of that high school course, nobody can see atoms. Using this mode of thinking, chemistry is purely cerebral and we use our eyes only to detect chemical changes. I have since learned, however, that I was a bit naive and much of chemistry relies on vision purely to make the science efficient. That is, conveying chemical data between colleagues, collaborators, or to students is customarily done using visual means because vision is such an efficient sense to have.

My high school instructor became a true ally and worked with me to figure out how to make her class accessible to me. The details of the modifications we made will be expressed a bit later but for now, suffice it to say that I worked with an assistant to carry out laboratory work. This assistant was one year ahead of me and had excelled in the course the previous year. This experience of working with a competent assistant gave me the opportunity to learn how to work productively with assistants and how I wished to be treated as a student.

Success in my honor's chemistry class gave me the confidence to enroll at the University of California, Davis and pursue a chemistry major. Much of pursuing a collegiate chemistry major for a blind person involves gently teaching the chemistry department that we can indeed think about chemistry independently and that we can handle finding, training, and working with assistants. The problem arises when students who are BVI are not yet ready to gracefully take on this gentle educational role. Much of my early years in the UC Davis chemistry department involved my reassuring people that I would find assistance and that the chemistry department need not worry about my success.

My time as an undergraduate student was an intense learning experience for me; I had to learn how to balance my work and my life while working very closely with assistants in various subjects. For all non-laboratory work, I worked mostly with a reader and braille transcriber who helped me create an extraordinarily productive working environment. My main reader did everything she could to ensure that materials were presented to me in an accessible format. She made it possible for me, as a blind student, to focus my time with new or less experienced assistants for laboratory work.

The science laboratory is a fast-paced and visual place where safety is paramount. As such, I required an assistant to be with me whenever I was in the experimental chemistry lab. It took several quarters of trial and error before I established a set of best practices when working with assistants in the lab. By meeting with assistants for several hours prior to the lab period, I was able to fully read and write down the procedure and discuss with my assistant precisely what each of our roles was. The idea behind these meetings was two-fold. First, I now had access to the printed procedure through my assistant so I understood it at the same level as my assistant. Secondly, using this pre-meeting method, I could be sure that my assistant actually took the time to read the procedure themselves rather than telling me they would read it and coming to the lab unprepared. I learned this lesson through experience; by having assistants fail to read the procedure and then arrive to lab not knowing how to do the lab exercise and being completely unprepared when it came to safety.

The chemistry laboratory can be a safety hazard if not treated properly. It must be treated with great respect. As the student enrolled in the lab class, I realized that I am responsible for both my safety and my assistants' safety. In other words, I took the responsibility of making sure my assistant knew precisely what I needed in terms of safety and I always made sure that he or she understood the potential dangers of the experiment at hand. Some of my lab assistants were not chemists by training and required extra careful discussion prior to lab. This is a third reason why I believe the somewhat intense pre-lab meetings between students with disabilities and assistant are indispensable.

Meeting with assistants before lab also makes the actual lab sessions quite straightforward. My assistant and I have a plan that we can execute, he or she knows that I want to play an active role in the experiment and do as much of the work myself as possible, and we execute the lab. Pre-lab meetings also increased both of our efficiencies so we could get through lab exercises faster. While in the lab itself, the student-assistant relationship needs to be fine-tuned so that the student is directing the activities in lab while the assistant is merely acting as the student's eyes. I have observed numerous examples of BVI students finding competent assistants, bringing them to lab, and letting the assistant do all of the work with little to no communication with the student. This gives the student who is BVI the opportunity to be lazy and not do the work required. Students learn quickly in college that this technique does not work because the student needs to learn the material in order to do well on exams. Ultimately, I prefer a relationship where I am in constant verbal communication with my assistant, I know precisely what he or she is doing and he or she knows exactly what I am doing (*12*).

I do not expect my assistants to be watching me all the time. As such, whenever I take action in lab, I tell them what I am doing with as much verbal description as I want them to use when they tell me what they are doing. This way, they know what I'm doing without watching me and it reminds the assistant what level of communication I need to succeed.

Assistants are most efficient when they have some knowledge of the subject matter and have taken the course beforehand. This is in no way necessary but it does make the pre-lab process faster and makes the overall work flow more efficient. I have run into problems with very experienced assistants who try to do everything for me and, in a sense, push me out of the experiment space. Therefore, we have a conversation beforehand where I make sure that my assistants know that I want to be as much a part of lab as they are. This can be a difficult balance between supervising, telling assistants what I need and also maintaining a suitable professional rapport. I often teach students who are BVI how I like to work with assistants because many students run into the problem of the assistants feeling domineered or controlled. Just like any working situation, anyone feeling controlled is never productive and both parties need to work collaboratively to keep the professional relationship strong and productive.

I also learned over time that meeting with the particular assistant with whom I performed the lab after the given specific lab session to tabulate data is useful. If I work with a different assistant to process data, they need to be brought up to speed and the process becomes less efficient and more time consuming. Getting help organizing data, formatting laboratory reports, etc. takes time. In total, my pre-lab meetings typically run for about two and a half hours, lab sessions are typically three hours and the post-lab sessions tend to run about three and a half hours for a total of nine hours per week of assistance per laboratory course. Now that I have this figured out, I can make the information clear to my assistants beforehand to ensure they are up for the task.

Prior to the authorship of this book, much of this information was not available to the public and students with disabilities are unaware of it. I spend a lot of time working with students with disabilities just beginning their undergraduate careers to teach them how I like to work with assistants and manage the collegiate workload. Many students have the attitude that they need to do everything for themselves and end up living an inefficient work life. Conversely, many students want to do nothing for themselves which leaves all their work up to their assistants. Then, their assistants are not as interested in helping because they do not see the students' initiative, the student does not learn what he or she is supposed to, and his or her grades reflect that.

I teach students that everything about life is a balance, including working with an assistant in the laboratory. They need to do their part but they need to ensure that their assistants are doing what they need as well. Finding assistants can be the hardest part of the job but once excellent assistants are found, I do everything I can to keep them interested in working with me. This message is difficult to teach other students with BVI. I have a genuine interest in performing excellent work and making performing excellent work exciting to my assistants. I also take a genuine interest in my assistants and what they are passionate about which engages them and motivates them to work with me. I have found that many students are really concerned with themselves, they do not take enough interest in their assistants and their assistants get bored working with them and leave. This is a problem that I try to help students avoid early on.

Ultimately, more students with disabilities should feel comfortable about engaging in science, technology, engineering and math (STEM) fields of study. I founded Accessible Science, a nonprofit organization to demonstrate this message to students (*13*). If we have students with disabilities joining STEM fields, they need to be ready to do what it takes to be safe and practical in the laboratory. This was a one main motivation for me to publish in the Journal of Chemical Health and Safety (*14*). The work presented focuses on safety in the high school and undergraduate laboratory spaces as well as work I have embarked on to train and excite sighted assistants. I hope this book reaches the scientific community broadly and shows students that it is not impossible to be a scientist with a disability. I also hope it demonstrates to them that they must take appropriate actions for both their own and their assistants' safety and that they must take responsibility for their education.

We have completed several other projects related to making chemistry, through bench work and computational, more accessible to students with disabilities (*15–17*).

Conclusion

Persons with disabilities face unique challenges that individuals without disabilities do not face. Social and personal preconceptions about limited abilities of persons with visual or hearing disabilities often discourage individuals with these disabilities from studying fields or disciplines that seemingly rely on having all five senses intact. Science, technology, engineering and math (STEM) disciplines certainly fall in this category and laboratory sciences like experimental chemistry are among the most discouraged. This is perhaps because it is believed that people who lack the ability to see or hear perfectly will be more a risk of danger in the laboratory.

Herein, we attempted to demonstrate how students with visual and/or hearing disabilities can safely and successfully master making the experimental laboratory a safe and productive place for them to perform lab work. Certainly assistance is highly recommended and the method of making the laboratory safe and productive depends largely on the student's personal disabilities and learning styles.

Ultimately, we believe that lack of vision or hearing is merely an inefficiency. If one has a visual or hearing disability, he or she just needs to figure out the best strategies for he or she to succeed. Anything is possible with adaptation, collaboration, and diligent work ethic. If those of us with vision or hearing disabilities take responsibility for ourselves and how we learn, work with others to ensure our colleagues or instructors are comfortable with our disability, and acknowledge that laboratory work will take us longer, we can all excel safely and productively in the experimental laboratory.

References

1. Erickson, W.; Lee, C.; von Schrader, S. *Disability Statistics from the American Community Survey (ACS)*. Ithaca, NY: Cornell University Yang-Tan Institute (YTI). 2017. www.disabilitystatistics.org (accessed November 27, 2017).
2. Legal Information Institute, Cornell Law School. 42 CFR §12102. https://www.law.cornell.edu/uscode/text/42/12102.
3. American Foundation for the Blind, Key Definitions of Statistical Terms. 2017. http://www.afb.org/info/blindness-statistics/key-definitions-of-statistical-terms/25 (accessed November 27, 2017).
4. Loy, B. Job Accommodation and Network Compliance Series: Employees with Vision Impairments. 2013. https://askjan.org/media/Sight.html (accessed November 27, 2017).
5. Burgstahler, S. Making Science Labs Accessible to Students with Disabilities. 2012. http://www.washington.edu/doit/making-science-labs-accessible-students-disabilities (accessed November 27, 2017).
6. National Association of the Deaf, Community and Culture, Frequently Asked Questions. https://www.nad.org/resources/american-sign-language/community-and-culture-frequently-asked-questions/ (accessed November 27, 2017).
7. Padden, C. A.; Humphries, C. *Deaf in America; Voices from a Culture*; Harvard University Press: Cambridge, MA, 1999.
8. Center for Parent Information Resources, Deafness and Hearing Loss. Disability Fact Sheet 3 (FS3), 2010. http://www.parentcenterhub.org/hearingloss/ (accessed November 27, 2017).
9. MacDonald, G.; Seal, B. C.; Wynne, D. H. Deaf Students, Teachers, and Interpreters in the Chemistry Lab. *J. Chem. Educ.* **2002**, *79*, 239.
10. Susan, S. B.; Ross, A. D.; Pagano, T. Chemical and biological research with deaf and hard-of-hearing students and professionals: Ensuring a safe and successful laboratory environment. *J. Chem. Health Saf.* **2016**, *23*, 24–31.
11. Burgstahler, S. The Winning Equation: Access + Attitude = Success in Math and Science. 2012. http://www.washington.edu/doit/winning-equation-access-attitude-success-math-and-science (accessed November 27, 2017).

12. Wedler, H.. Nobody Can See Atoms: A Comprehensive Discussion of Computational Chemical Results and Making Chemistry Accessible to All. Ph.D. Thesis, University of California, Davis, Davis, CA, December 2016.
13. Accessible Science. http://www.accessiblescience.org/ (accessed November 27, 2017).
14. Nepomuceno, G. M.; Decker, D. M.; Shaw, J. D.; Boyes, L.; Tantillo, D. J.; Wedler, H. B. The Value of Safety and Practicality: Recommendations for Training Disabled Students in the Sciences with a Focus on Blind and Visually Impaired Students in Chemistry Laboratories. *J. Chem. Health Saf.* **2016**, *23*, 5–11.
15. Wedler, H.; Cohen, S. R.; Davis, R. L.; Harrison, J. G.; Siebert, M. R.; Willenbring, D. ; Hamann, C. S.; Shaw, J. T.; Tantillo, D. J. Applied Computational Chemistry for the Blind and Visually Impaired. *J. Chem. Educ.* **2012**, *89*, 1400–1404.
16. Wedler, H.; Boyes, L.; Davis, R. L.; Flynn, D.; Franz, A.; Hamann, C. S.; Harrison, J. G.; Lodewyk, M. W.; Milinkevich, K. A.; Shaw, J. T.; Tantillo, D. J.; Wang, S. C. Nobody Can See Atoms: Science Camps Highlighting Approaches for Making Chemistry Accessible to Blind and Visually Impaired Students. *J. Chem. Educ.* **2014**, *91*, 188–194.
17. Lounnas, V.; Wedler, H. B.; Newman, T.; Schaftenaar, G.; Nepomuceno, G.; Pemberton, R. P.; Vriend, G.; Tantillo, D. J. Visually Impaired Researchers Get their Hands on Quantum Chemistry; Application to a Computational Study on the Isomerization of a Sterol. *J. Comput.-Aided Mole. Design* **2014**, *28*, 1057–1067.

Chapter 7

Service Dogs in the Chemistry Laboratory

Patricia Ann Redden*,[1] and Christopher Scott Sweet[2]

[1]Chemistry Department, Saint Peter's University, 2641 Kennedy Blvd., Jersey City, New Jersey 07306, United States
[2]K. Lisa Yang and Hock E Tan Institute on Employment and Disability, Industrial and Labor Relations School, Cornell University, 312 Dolgen Hall, Ithaca, New York 14853, United States
***E-mail: predden@saintpeters.edu**

It is becoming increasingly common to encounter individuals accompanied by service dogs in all aspects of our daily lives, and that means that the question of access by service dogs into schools and laboratories is becoming more important. While an individual with a disability has the right under the American with Disabilities Act (ADA) to be accompanied by a service animal, there are a number of practical considerations in making that access safe for the animal, its partner, and other individuals in the laboratory. In this chapter, we will attempt to identify these considerations and suggest procedures to be followed.

As the use of service dogs becomes more prevalent in society, it is increasingly likely that students, faculty members, and working chemists will seek to bring their dogs into the laboratory setting. Although such access is allowed under the Americans with Disabilities Act (ADA), the laboratory setting necessarily introduces special concerns about safety for the individual with the disability, the dog, and other occupants of the laboratory. This chapter will discuss the regulations that govern access as well as the practical considerations that the laboratory supervisor should discuss with the dog's partner to address the safety concerns.

Definition and Rights of a Service Animal

Titles II and III of the Americans with Disabilities Act (ADA), as revised in 2011, define service animals as dogs that have been specifically trained to work with an individual with a disability, performing a task for that individual that mitigates the effect of the disability (*1*). In some limited situations, the ADA also identifies miniature horses as alternatives to service dogs, although they are not included in the definition of a "service animal." Miniature horses generally have a maximum height of 35 inches at the shoulder, weigh up to 100 pounds and have been used primarily for individuals with motor or visual impairments. The horse must be housebroken and the facility would have to be able to accommodate the horse's size, type and weight. Since their use is limited, they will not be considered further in this chapter.

The task performed by the service dog must be related to the disability, which may be physical, sensory, psychiatric, intellectual or other mental disability. The most well-known examples of service dogs are guide dogs for the visually impaired, but a dog may also provide support for an individual with a motor impairment or hearing loss. The dog may work with an individual who has seizures or diabetes, identifying imminent medical problems, or may help a person in a wheelchair by picking up or handing off items, opening doors, even providing help in dressing. Some dogs are trained to work with a partner with Post-Traumatic Stress Disorder (PTSD) who is subject to anxiety attacks, with a partner who has autism, or with individuals with other neurological conditions such as traumatic brain injuries. The key is that the dog must be able to perform a task that is directly related to the partner's disability, not simply provide comfort. The dog may be of any size or breed, literally from a Chihuahua to a Great Dane. It may have been obtained from a service dog organization such as The Seeing Eye or Canine Companions for Independence (two of the oldest and largest national organizations), or it may have been trained by the individual partner. In some cases, an individual with multiple disabilities may actually have more than one service dog, each trained to respond to a different need (*2*).

Often confused with service dogs are two other categories, emotional support animals and therapy animals. "Emotional support animals" are not only dogs but may literally be any type of animal, including but not limited to cats, miniature pigs, snakes, and birds. These animals provide comfort to the partner but don't perform a specific task related to the disability. They do not have public access rights, although they may be permitted in housing (including college residence halls) and on domestic and international air travel. Legislation that does permit this type of animal includes the Federal Fair Housing Act and the Air Carrier Accessibility Act. "Therapy animals" typically interact with a number of individuals in either a public or private setting; examples of public settings include hospitals, rehabilitation centers, libraries, or educational settings during high stress periods, such as exam week. Again, therapy animals do not have public access rights and may include other types of animals (*3*).

The ADA requires that state and local government agencies, businesses, and non-profit organizations that provide employment, goods, or services to the public must make "reasonable accommodations" for individuals with disabilities to make it easier for that individual to access the service or to perform the duties of the position, unless that accommodation would cause significant difficulty or expense (*4*, *5*). This requirement includes public assess for service dogs. As a result, service dogs are allowed in all areas where the public is allowed; although religious institutions and organizations are specifically exempt from the ADA since they are not considered to be public accommodations. Access includes, but is not limited to, hotels, public transportation, offices, educational settings, stores, restaurants, and theaters. Additionally, nonreligious entities that conduct their activities in church facilities may also be subject to the act (*4*). Service dogs may only be excluded if they are out of control, if they are not housebroken, if their presence would "fundamentally alter" the nature of the service or program, or if it would violate "legitimate safety requirements" (*3*, *6*).

It is important to note that in employment settings under Title I of the ADA, as enforced by the Equal Employment Opportunity Commission (EEOC), service animals are not automatically provided access, although employers have an obligation to consider the use of a service animal as a form of reasonable accommodation (*5*). In cases where the person's disability is not obvious, employers may ask for documentation describing the employee's functional limitations and outlining the specific need for the animal in performing the essential job functions. Service animals may be excluded from the workplace if they pose either an undue hardship or a direct threat in the workplace (*2*). However, the presence of other individuals with allergies or fear of dogs is not an adequate reason for allowing access (*1*).

As the number of legitimate service dogs increases, so does the incidence of "fake" service dogs. These are typically dogs whose owners may honestly misunderstand the definition of a service dog (confusing them with emotional support or therapy dogs) or who deliberately want to bring pets into situations normally not permitted. This has resulted in situations where legitimate service dog teams are harassed, distracted or barred by a venue because of experiences with uncontrolled, non-housebroken, or poorly behaved animals. To exacerbate the problem, it is possible to purchase vests and "service dog" identification cards online with no check on legitimacy. In fact, contrary to popular opinion, neither proof of specialized training, vests, special harnesses, or identification is needed for a service dog, although many do have these. Also complicating the issue is that, although most people would think of service dogs as German Shepherds, Labrador Retrievers, or Golden Retrievers, any dog might be a service dog.

In addition to trained service dogs, requests for access may be made in some cases for puppies in training to become service dogs. Puppies bred by service dog organizations are usually raised and socialized for up to eighteen months by individual volunteer puppy raisers, then returned to the organization for the professional "task" training. The puppy raisers and trainers are expected to expose the puppies to as many situations as possible, to prepare them for the future. Many of these puppies are therefore brought to the same types of venues as a graduate service dog. However, unless allowed by state or local law, or by individual venues, these puppies do not have public access rights, although the venue may allow access. Some states, New York for example, do allow service dog trainers engaged in the actual training of these puppies to have the same rights and privileges with respect to access to public facilities as working service dogs.

How, then, can an instructor or supervisor determine if a particular animal is really a service dog and entitled to public access? Some situations would be obvious: a guide dog for an individual with a visual disability wearing a harness with a handle or a service dog accompanying an individual in a wheelchair, possibly helping to pull it. If it is not obvious that the dog is a service dog, only two questions may be asked of the partner:

(1) Is the dog a service animal required because of a disability?
(2) What work or task has the dog been trained to perform?

The person asking the questions may not ask the nature of the partner's disability, nor can documentation/identification of the dog as a service dog be requested. Finally, the partner cannot be asked to demonstrate the task (*2*).

Once the dog is identified as a service dog, it can only be denied access or removed from the premises if it is not housebroken, is a threat to the health and safety of others, or is out of control and its partner (or an accompanying aide) does not take effective action to control it. The presence in the room of other individuals with allergies or fear of dogs is not a valid reason for restricting access by the service dog team. Normally, the dog must be harnessed, leashed or tethered, but in certain circumstances the dog may be controlled by voice, signal or other means (*1*). For example, a diabetic alert dog may typically be carried in its partner's arms or a chest pack to be close enough to its partner's mouth to recognize changing glucose levels. An individual with PTSD may have a dog trained to enter unfamiliar spaces to check that there is no threat, and therefore the dog must be off-leash to fulfill that task.

Access to Laboratory Space

Given that service dogs do have access rights to public areas as described above, the question now is specifically how to deal with them in laboratory settings, both in academic institutions and in industrial settings. There are real safety concerns for the dog, its human partner, and other individuals in the laboratory. These concerns must be balanced with the needs of the human partner for the dog's services and with the law. Many academic institutions have established policies for service dogs, and in some cases for service dogs in training, on their campuses and in their laboratories, and these policies can usually be found on their websites. In most cases, these policies identify laboratories, as well as other areas where protective clothing is worn, as off-limits, although they will usually allow access on a case-by-case basis or with protective equipment. An example of such a comprehensive policy can be found on the website of Westminster College (*7*). The allowance for an exception on an individual basis puts the onus on the laboratory director to evaluate the situation and make a decision, with input from the dog's partner. Points that should be considered are: the need of the partner for the service dog in the laboratory; the types of chemicals used and chemical operations conducted; the physical layout of the laboratory; and protection for the dog.

Need for the Dog's Services

The partnership with a service dog increases the independence of the partner and decreases dependence on others. Barring a dog from the laboratory negates these benefits. However, a realistic evaluation of the task(s) the dog performs is important. A service dog for an individual who uses a wheelchair may be trained to pick up items from the floor and bring them to the partner. This is obviously something that no one would want the dog to do in the laboratory setting because of the possibility of contamination and harm to the dog, even if the item being picked up is simply a pen. However, another such individual might have a dog trained to carry a bag or basket with needed materials or might depend on the dog to pull the wheelchair. A hearing dog might alert its partner to a fire alarm or other sound indicating an emergency, but it might not be able to discriminate between these sounds and more normal laboratory sounds like vacuum pumps or laboratory chemical hood operation. In these examples, the partner may be willing to leave the dog out of the lab and work with a colleague to carry out the needed tasks. However, a dog which supports balance, alerts to an incipient seizure or is trained to warn of a change in glucose levels might be required at all times.

Another consideration is the manner in which a dog alerts its partner. Most hearing, seizure alert and medical alert dogs nudge the partner's leg or arm in a situation they feel needs a response, but other dogs may alert by jumping or barking. Either method could create a hazardous situation if the partner is handling chemicals.

Chemicals and Chemical Operations

It goes without saying that laboratories run the gamut from individual research laboratories to large academic or industrial laboratories, from wet chemistry to instrumental laboratories, dealing with organic, inorganic or biochemical materials. Each of these presents a unique challenge.

Although it is to be hoped that adequate laboratory chemical hoods are available, there may be situations with less-than-adequate ventilation or hood space. In that case, heavier-than-air vapors can sink to and travel along the floor. Since a dog's nose is far more sensitive than a human's, even small quantities of vapors can be potentially irritating or harmful. Safety Data Sheets (SDS's) give information about a chemical's effect on humans, but dogs may be more or less sensitive than humans, and that information may be hard to obtain.

Small amounts of solids or liquids may drop from lab benches to the floor or on to the dog, again presenting a potential hazard to the dog, and even cleanup of such spills may leave residue on the floor. Student laboratories, particularly undergraduate ones, are typically very prone to spills, drips, and dropped glassware. Containers for chemical waste may be on the floor, as well as carboys for solvents.

An instrument laboratory, on the other hand, may expose the dog to fewer chemicals but may have other potential problems. A compressor or vacuum pump generates noise, which can upset or startle the dog and cause unexpected movement on its part. Intermittent or continuous background noise could distract the dog from its intended task, as in the case of a seizure alert dog that needs to be cognizant of very small changes in its partner's behavior. Some instrument components might be on the floor rather than on a bench, minimizing the free space for movement or placement of the dog.

Physical Layout of the Laboratory

Just as chemical operations vary significantly among laboratories and must be considered, the amount of open floor space and the items that may be on the laboratory floor are important. Consider the floor area of the laboratory, the number of people working in the laboratory, and the arrangement of the benches. Are there solvent or waste containers, pumps, glassware, or chemicals on the floor or on low open shelving? Where are the normal high traffic areas, showers and eyewash units, and emergency exits?

Students in an academic laboratory are warned against leaving backpacks on the floor or drawers open, to avoid tripping hazards, and for the same reason laboratory stools are usually not allowed in academic laboratories, except in balance rooms or instrument areas. A service dog next to a bench presents a similar tripping hazard, one potentially exacerbated by a tail or legs protruding into the aisle. A service dog typically is trained to lie quietly in one location until it is needed. However, even the best-trained dog will shift position over the course of several hours, which could create a sudden tripping hazard. Areas of a laboratory that are more heavily travelled than others should also be avoided for the dog and its partner.

The best options for where to locate the dog would include spaces where the dog will not impede traffic but will be near its partner. A "wheelchair recess" area under a laboratory bench would potentially be a better place for the dog than the aisle, even if the space is shared with the human partner; a corner of the laboratory would be another possibility. In some cases, dogs have been placed under coat racks or balance tables, even in crates in a corner of the laboratory.

If the decision is made that the dog is not needed in the laboratory, it may be left outside the laboratory or in a location away from the partner. In that case, it should be placed in a safe, comfortable place where it will be protected from and not disturbed by passers-by and where it can be quickly retrieved in case of an emergency or when the chemist leaves the laboratory.

Protection for the Dog

Consider personal protection equipment and dress required for a student in a chemistry laboratory, promulgated by the ACS Committee on Chemical Safety in its publication, Safety in Academic Chemistry Laboratories, 8th edition (*8*). Recommended dress includes close-toed shoes made of leather or leather-like substitutes and a chemical- and fire-resistant lab coat or apron, and the amount of exposed skin should be minimized. Hair should be tied back, loose sleeves confined, and all jewelry removed. Eyes are protected with chemical splash goggles and, if necessary, face shields. Appropriate gloves are used to protect the hands.

Similar consideration should be given to the service dog in the laboratory. Since its entire body is below bench level, it is subject to drips, spills and splashes of chemicals. It walks and lies on the floor, which may have broken shards of glass or spilled chemicals. Heavier-than-air vapors may concentrate on the floor area. For full protection, exposure by all these routes of entry should be considered. Goggles will protect the eyes and booties the feet, and the dog may be given a mat to lie on. However, its entire body is also exposed, including head, back, underbody and tail. Additionally, a dog will lick its fur, particularly if it gets wet, which would allow it to ingest chemicals. Figure 1 shows a service dog at the University of Colorado, Denver, wearing swim goggles and booties, sitting on a mat, with minimal protection for its body.

Figure 1. Service dog with minimal protection (9)

Some individuals and companies have made full body suits for the dog from Tyvek or other materials to give protection from contact, spilled and ingested chemicals. A secondary advantage to such suits is that they also protect the laboratory environment from fur, dander and bacteria that could cause contamination. One such suit, the Canine Cleanroom suit developed by Industrial Safety Professionals, Inc., is shown in Figure 2. The suit was designed for protection for dogs trained to detect molds, but it could also be used for protection from chemicals. This type of suit must be fitted to the dog and the dog must be evaluated to see if the suit interferes with the dog's tasks.

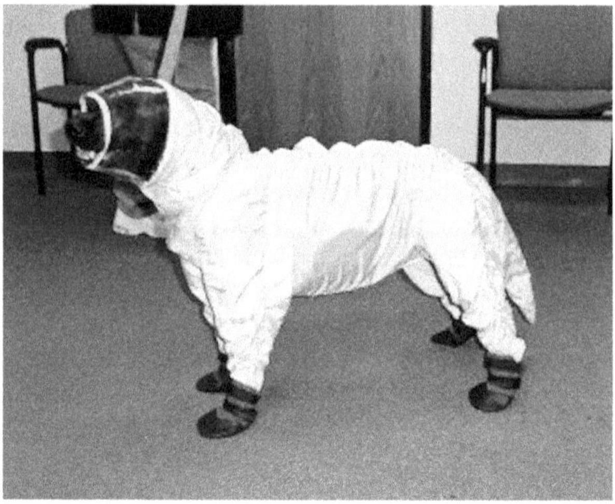

Figure 2. Canine Cleanroom suit for complete protection (10)

Institutional or Company Policies

It would be advisable to have a written policy for service dogs, preferably one that requires discussion among the individual with the service dog, the laboratory supervisor(s), safety officials and possibly a representative from Human Resources. As noted earlier, the written policy may indicate that dogs are not normally allowed in certain areas but allow for exceptions. Documenting how such exceptions will be identified is valuable to ensuring the institutional policies are in compliance with the Americans with Disabilities Act, the Federal Fair Housing Act, and the Air Carrier Accessibility Act. For specific cases, a summary of the discussions and of the accommodations should be in writing, signed by all parties, and kept as a permanent record.

Ideally there should also be a process for alerting security personnel to the presence of the dog in the laboratory, in case an emergency occurs in the facility. This process would be especially important in the case of a medical emergency involving the dog's partner, when security might have to interact with the dog in a stressful situation. An identity badge or tag for the service dog could be used to show that its presence has been discussed and approved (*3*).

Conclusion

The Americans with Disabilities Act gives individuals with disabilities the right to bring their service dogs into public areas, including laboratories, with reasonable accommodations. However, a chemistry laboratory does expose the dog to some unique and potentially hazardous situations, and a decision to bring the dog into the laboratory must include an identification of potential hazards and methods of mitigating them. The dog's partner and supervisors should consider the areas discussed in this chapter, so that an informed decision can be made on its access to the laboratory. This is in the best interests of the service dog and of its partner, as well as of others working in the laboratory.

References

1. ADA Requirements; U.S. Department of Justice, Civil Rights Division, Disability Rights Section. https://www.ada.gov/service_animals_2010.htm (accessed July 28, 2017).
2. Service Animals and Emotional Support Animals; ADA National Network: Information, Guidance and Training on the Americans with Disabilities Act, Resources. https://adata.org/publication/service-animals-booklet (accessed July 28, 2017).
3. Frequently Asked Questions about Service Animals and the ADA; U.S. Department of Justice, Civil Rights Division, Disability Rights Section. https://www.ada.gov/regs2010/service_animal_qa.html#bre (accessed July 28, 2017).
4. Disability Discrimination: U.S. Equal Employment Opportunity Commission, Laws, Regulations and Guidance. https://www.eeoc.gov/laws/types/disability.cfm (accessed July 28, 2017).
5. Appendix to Part 1630, Interpretive Guidance on Title I of the Americans with Disabilities Act; Cornell Law School, Legal Information Institute. https://www.law.cornell.edu/cfr/text/29/appendix-to_part_1630 (accessed July 28, 2017).
6. Service Animals; ADA National Network: Information, Guidance and Training on the Americans with Disabilities Act, Resources. https://adata.org/factsheet/service-animals (accessed July 28, 2017).
7. Policies and Procedures for Service Animal: Westminster College Disability Services. https://www.westminstercollege.edu/docs/default-source/departments-and-offices-documents/disability-services-documents/policies/policies/service-animal-policy.pdf?sfvrsn=4 (accessed July 28, 2017).
8. American Chemical Society Joint Board-Council Committee on Chemical Safety. *Safety in Academic Chemistry Laboratories*, 8th ed.; American Chemical Society: Washington, DC, 2017; pp 16−18.
9. Garcia, M. Puppy Protective Equipment, Retrosynthetic Rummy. *Chem. Eng. News* **2015**, *93* (3), 48; http://cen.acs.org/articles/93/i3/Puppy-Protective-Equipment-Retrosynthetic-Rummy.html (accessed July 28, 2017).
10. Canine Cleanroom Suit | Inventor Bradley A. Prill, CIH, CSP, holds Utility and Design Patent; Industrial Safety Professionals, Inc., 27475 Ynez Rd, Suite 376, Temecula, CA 92591; www.indspi.com (accessed July 28, 2017).

Chapter 8

Laboratory Safety for All: Accommodating Students with Disabilities in Chemistry Teaching Laboratories

Joanna Boval[1] and Sheila Kennedy[2,*]

[1]Office for Students with Disabilities, University of California, San Diego, 9500 Gilman Dr. MC0019, La Jolla, California 92093-0019, United States
[2]Teaching Laboratories, Chemistry & Biochemistry Department, University of California, San Diego, 9500 Gilman Dr. MC0303, La Jolla, California 92093-0303, United States
[*]E-mail: s1kennedy@ucsd.edu

As more students with disabilities study STEM fields, universities must re-examine laboratory programs and environments to ensure accessibility. At UC San Diego, the Office for Students with Disabilities and the Chemistry & Biochemistry Teaching Laboratories work with the faculty, instructors and teaching assistants to meet students' needs while maintaining the health and safety of all laboratory participants.

Academically qualified students come into chemistry laboratory courses with a variety of abilities and challenges. Some arrive with chronic health issues; some have physical limitations such as low vision, restricted reach or grasp strength; some have mobility issues. At the University of California, San Diego (UC San Diego), the Department of Chemistry & Biochemistry (CHEM) and the Office for Students with Disabilities (OSD) work with the faculty, instructors and teaching assistants to provide an accessible laboratory experience for all students. Our best success stems from out-of-the-box thinking, collaboration and adherence to a few basic guidelines.

Legislative Mandate

Federal law, the Americans with Disabilities Act (ADA; 1990, amended 2008), requires that a "qualified individual with a disability" receive "reasonable" accommodation in public entities and in programs or activities receiving federal financial assistance (*1*, *2*). Programs to ensure such reasonable accommodations for workers and students with varied abilities are often called "ADA programs."

Many university campuses have a staff group that is trained to support students with disabilities and ensure that accommodations are made available. It may be called an "Office for Students with Disabilities," "Disabled Student Services" or "Disabled Student Services Program." The professionals in this office are the experts on disability for their campus. The specialists in this office know what is required by law (Federal and State), what is available on their campus and in the community, and whether funding is available for the purchase of specialized instruments, tools and furniture.

Implementation at UC San Diego

UC San Diego is a major research university with over 7600 graduate students and more than 28,000 undergraduates. The campus operates on a three-quarter (nine-month) academic year. Like many departments, CHEM offers a full slate of courses, including laboratory courses, during Summer Session.

At UC San Diego, OSD guides faculty, staff and students in finding resources and in defining a "reasonable accommodation." OSD employs four full-time Disability Counselors with backgrounds in psychology, counseling, social service and special education. In addition, they receive specialized on-going training.

OSD works with about 500 students each quarter, reviewing documents from medical professionals and issuing Authorization for Accommodation (AFA) letters to the faculty. Each AFA letter is issued for a specific 10-week quarter, since the limitation of the disability may impact the student and his/her studies in different ways in different environments or courses.

In our experience at UC San Diego, more than half of the students receiving accommodations have psychological disabilities, such as a generalized anxiety disorder or major depression. While many college students have a high level of anxiety, not all reach the level of "disability." College students, however, are at the prime age for new diagnoses of depression, schizophrenia and bipolar disorder (*3*).

After those with psychological disabilities, the next largest group is students with chronic health issues. Many of these are well managed, but some are newly diagnosed. The impact of a disability may wax and wane as the student experiences periods of flare and remission. Learning disabilities, attention deficit–hyperactivity disorder (ADHD) and autism spectrum disorder appear with smaller numbers, followed by recent, temporary injuries, such as broken or sprained legs, ankles, wrists and shoulders.

CHEM is a founding department at UC San Diego and has a major research component with 63 faculty members, 140 Postdoctoral and Research Scholars and

269 graduate students. CHEM has one of the largest undergraduate programs in the country with a growing population of approximately 1000 major students plus a service population of roughly 21,500 from other departments. Including those students from other departments, our instructors and staff see just over 4000 students each academic year enrolled in more than 180 sections of laboratory courses.

Students normally register for new courses, then go on a school break for a week or two. When some of them return to their laboratory course with broken bones and sprained ankles or in the process of recovering from recent surgery, the course instructors and laboratory staff make accommodations, find ways to include them in laboratory activities and keep their educational plans on track, where ever possible.

Guiding Principles

"Where ever possible" can become a difficult balancing act when department instructors and staff work with OSD to consider what is possible and what is required to facilitate a student's participation in a laboratory course. Our discussions with instructors and staff members across campus have revealed a clear set of priorities for students working in laboratories. We have shared these with our colleagues (*4*, *5*) and included them in our article on adaptive laboratory tools in the Journal of Chemical Health & Safety (*6*). We quote them again below.

In our work finding accommodations, we have developed a routine of consultation among the student, the instructor, the laboratory teaching assistant and OSD. Discussion is kept focused on the student's functional limitations, while trying to answer, "What specific laboratory task(s) does the student find difficult?" We do not inquire about the student's medical history or diagnosis, although some students do share this information.

GUIDING PRINCIPLES (*6*)

1. ***No one gets hurt.*** *A student who cannot, with reasonable accommodations, work in a way that guards his/her own safety, as well as that of the others working in the laboratory, should not be working in a laboratory. This applies to all students, not just those requesting special assistance.*
2. ***Accommodate everyone*** *who wants to learn laboratory skills, if at all possible.*
3. ***Reasonable accommodation*** *may include considerations of scheduling. A student whose broken leg confines him to a cast and crutches for six weeks is encouraged to reschedule the laboratory course for another quarter, where the pain and the mobility issues will not be factors. The student is reminded that he will have a better chance to give the course his full attention and perform at his best. For a student with a permanent*

disability, delaying enrollment would serve no purpose, but time may be needed to acquire adapted tools or furnishings.

4. ***Laboratory credit for laboratory work.*** *Curriculum and safety professionals across the campus agree that watching someone work is not the same as working. Laboratory course credit is given to students who have completed laboratory work. A student who has trouble manipulating glassware can be given modified glassware, extra time or extra tools to assist, but she must experience the work and make the decisions.*

Brainstorming and a willingness to think outside our usual paths can turn a difficult situation into an interesting puzzle to be solved by the team. Often the most valuable member of that team is the enrolled student. Living with a permanent disability, especially a long-established one, builds a wealth of problem-solving experience about everyday living. Sometimes we just need to apply the student's everyday solutions to the laboratory environment.

An additional consideration we have learned is using "person-first" language when speaking about and to people with disabilities. With some thought and practice, we have learned to speak of the person before speaking of that person's condition or disability. When changing established speech patterns (such as choosing to say "a student with a disability" rather than "a disabled student"), it helps to recall that we have more in common than we have differences.

The Accommodation Process

The process of providing accommodations works best when it is interactive and recursive. Faculty and staff recognize that a student with disabilities has met the same admission standards as all other students. Such students are found throughout UC San Diego's academic programs, departments and professional schools.

Students in primary and secondary education work under a different set of standards, defined in Section 504 of the Vocational Rehabilitation Act of 1973 and the Individuals with Disabilities Education Act (IDEA, 2004) (*2*, *7*). Students who have received assistance under IDEA may have experienced both accommodations and modifications to the curriculum (*7*).

- An **accommodation** changes the way a student accesses the curriculum.
- A **modification** of curriculum might include alternative educational activities and waivers of course requirements.

A student transitioning to university work learns that accommodations can be made, but modifications cannot. Course requirements will not be adjusted. Course work must be completed to get course credit and laboratory work is required for laboratory course credit.

Students may register with OSD as soon as they are accepted by the University. Those with previous experience of accommodations in high school or community college often take this route. OSD staff also attend orientation and recruiting events to contact new students and families.

The OSD Director maintains working relationships with directors of other campus offices, including Counseling and Psychological Services and Student Health Services. Both of these departments also refer students to OSD, as necessary.

The student provides documentation of a functional limitation (from a physician or other health care provider) and consults with a Disability Specialist from OSD. Together, they review the student's planned courses. The student may suggest useful accommodations, as accommodations that have worked in similar courses (in high school or college) may be used again. Department staff may be consulted about what activities a student needs to do to participate fully.

For a CHEM laboratory course, a Disability Specialist and the student meet with the Teaching Laboratories Safety Coordinator in the specific laboratory space so that the student can see and interact with the environment that will be used for the course. Observation of space and equipment that the student may find inaccessible leads conversation toward creating accessibility. Student input and feedback in this process is highly valued since students are more apt to commit to enrolling and remaining in the laboratory if they have a voice in the accommodation decisions.

OSD then issues an Authorization for Accommodation (AFA) to the student. The AFA requires the instructor and department to provide specific accommodations to the student for the specific course. Figure 1 is an example AFA.

Accommodations are made for relief or remedy to allow student access; no guarantee of good grades is made. The student is responsible for presenting the AFA to each instructor and requesting accommodations for each course. For the instructor, the AFA is a mandate that must be followed. If the AFA appears to represent a fundamental alteration to the course or program or if the requested accommodation would compromise health or safety, the instructor contacts OSD to request a revision to the AFA.

During the quarter, feedback is requested from both the student and the laboratory teaching assistant (TA) to assess whether changes are needed. Some students require – or benefit from – careful tracking, while others need very little.

Typical Accommodations

While we have learned to expect some of the more common requests for accommodations, we have also learned that students with disabilities vary as much as the student population as a whole. There is no checklist of items a student in a wheelchair will need. One student may be athletic, strong and full of energy while another works with a spectrum of limitations in addition to the mobility issue.

Each situation is evaluated according to what the individual needs to work safely and effectively on par with others in the course.

OSD Authorization For Accommodation (AFA) Letter
SP17

Issue Date: **6/8/2017**

To: Name of Professor (Course and Section)
Name of Department OSD Department Liaison

Re: **Name of Student and Student ID number**

This student has submitted medical documentation to OSD and is eligible for modifications, adjustments and other accommodations intended to minimize the impact of a documented disability. The following accommodations are required by the student's current functional limitations.

Note: The Americans with Disabilities Act (ADA) requires that universities provide reasonable and effective accommodations to students with documented disabilities. The faculty member is responsible for ensuring that these accommodations are fulfilled. In so doing faculty are not required to compromise the integrity of their courses.

<u>Classroom Accommodation (Facilitated Provided by OSD)</u>

- Note-taking (peer note-taking system)
- Alternate formats for required reading materials - Electronic format

<u>Classroom Accommodation (Facilitated Provided by Faculty or Faculty Designee)</u>

- Copies of handouts before class begins
- Copies of Power Point slides before class begins
- Extended time on labs (1.5 times the standard amount of time)
- Fine motor function and lab notebook entries-assistance in securing a scribe
- Lab partners: Permission to work alone or directly with a professor or TA
- Lab partners: assistance in securing
- Permission to type lab assignments
- Use of reserved stool in lab
- Use of respirator in lab

<u>Exam Accommodations (Facilitated Proctored by OSD)</u>

- Exam in alternate formats - Electronic format

<u>Exam Accommodations (Facilitated Proctored by Faculty or Faculty Designee)</u>

- Extended test time (1.5 times the standard amount of time; applies to in-class exams only and NOT to take home exams)
- Limit of one exam per day
- Separate, quiet, distraction-reduced testing (alone-no other students present)

The OSD, a department within the division of the Executive Vice Chancellor of Academic Affairs (EVCAA), is responsible for determining and facilitating UCSD accommodations for students with disabilities. The OSD is located at University Center 202 and our website is <u>http://disabilities.ucsd.edu/</u>. Questions or concerns regarding accommodations should be made to the OSD staff member listed below.

<u>Signatures</u>

Student Signature	Date	OSD Staff Signature	
Professor Signature	Date	OSD Liaison Signature	Date

Signatures of the faculty and/or OSD Liaison indicate that the AFA letter has been read, reviewed and discussed with the student. It does not imply agreement or disagreement with the contents of the letter.

Figure 1. Sample Authorization for Accommodation letter.

In a lecture course (including the lecture portion of a laboratory course), accommodations often include extra time on tests or assignments. Sign language interpreters and captionists may be brought into a lecture. Large-type materials may be provided. Some students receive lecture notes from instructors or have the lecture recorded. OSD recruits classmates to work as volunteer note takers and puts them in touch with students needing assistance.

Laboratory courses often have large lectures (300 – 400 students). The associated laboratory sections, with up to 24 students, are each taught by a graduate-level teaching assistant. Some of the laboratories are set up to partner students, while others require individual work. Most laboratory courses meet twice a week for three or four hours.

Laboratory accommodations may include extra time to complete laboratory work. In CHEM, this is arranged by reserving spaces in two back-to-back sections meeting in the same room. The student and the instructor usually need to coordinate with two teaching assistants. Extra time on exams may also be provided.

Time of day is important for some students who deal with complicated medication regimes or whose personal care leads to a slow start in the morning. Mornings are also difficult for students with multiple sclerosis, fibromyalgia and other chronic health issues. For these students, it's convenient that CHEM runs laboratory courses from 8:00am to 5:00pm.

Assistants may be used in laboratories. Sign language interpreters and captionists are assigned space in laboratory courses as well as lectures. The rule that everyone employed to work in a laboratory must complete the new employee safety training and follow all rules about personal protective equipment (PPE) applies to these assistants. When they are assigned to laboratories, CHEM provides appropriate PPE.

"Directed assistance" may be useful in a laboratory course where the enrolled student has physical limitations, such as grasp, reach or vision. In these cases, CHEM employs a student laboratory assistant to help the enrolled student. The assistant provides an extra pair of hands and is not to teach, coach, direct, or make decisions; the enrolled student must explicitly ask for help or direct the helper. The assistant may confirm an instrument reading but not read it for the enrolled student. Given the restrictive role of the assistant, it has worked better for the enrolled student and for the process if the assistant does not have previous experience with the course.

Physical spaces and equipment may be modified to accommodate students with disabilities. Most laboratories have sufficient space for wheelchair access, but table heights may need to be adjusted. A work space, including drawers that can be reached and access to utilities, including safety showers and eyewashes, may need to be arranged.

For some students, size and scale are important. A change to micro-scale glassware (especially in teaching organic chemistry) resulted in lower costs for glassware, chemical purchases and hazardous waste disposal fees. Returning to larger-scale models, instruments and glassware (sometimes found in storage) can make operations more accessible to those with low vision. For some students, magnification is useful.

For other students, transportation may be the major issue. At UC San Diego, personal cars on campus are actively discouraged and shuttle buses (available to all) make regular loops through the campus. In addition, an electric cart service provides transportation through pedestrian areas for those who cannot make the long walks that are common on campus. For long-term needs, riders may schedule service for weeks or for a whole quarter.

The student's need for personal protective equipment (PPE) may need to be re-evaluated based on circumstances that do not pertain to other students. Everyone in the course will work with the same materials and the hazards, being inherent in the material, do not change. Evaluation of the risk of injury, however, is based both on the hazard of the material and the likelihood of exposure to that hazard. Decisions about required PPE are based on this combined risk assessment. If a student's circumstances increase the proximity to the hazard, or make an otherwise-unlikely route of exposure more likely, then the student needs alternative or additional PPE. The primary example of this situation is a student who must work from a seated position. That student's lap is exposed to spills and drips more than that of a student who stands. A corrosive-resistant apron can be added to the normal lab coat to provide that extra protection.

When new challenges appear and when we have a chance to design or install laboratory equipment, it is often helpful to refer to the Principles of Universal Design (Figure 2), which direct us toward products and facilities that are designed for use by all, to the greatest extent possible, without adaptation (*8*). A wheelchair user may have been the intended beneficiary of laboratory workstations with lower counter heights, but they also function well for students who are short of stature. Likewise, lever-action handles on doors and faucets are easier for everyone, where round knobs are impossible for some. Automatic door openers help those using wheelchairs or walkers, but are also convenient for the elderly or those carrying heavy loads or small children. Ensuring adequate room for passage of a wheelchair provides room for all.

Case Studies

Some of the following case studies and photos have been previously published in our article on adaptive tools (*6*). The low-vison case study was presented at the symposium "Supporting and Facilitating Students and Workers with Disabilities in the Research and Teaching Laboratory" at the 248th National Meeting of the American Chemical Society, San Francisco, CA, 2014 (*5*).

Hand Grips

For students who have trouble grasping and controlling small tools, including pipets, pipettors, spatulas and syringes, we have increased the size of grips and handles. While these tools are commonly demonstrated using a single steady hand and strong small motor skills, we find increasing the grip diameter and planning to work with a two-hand grip allows a student with weak small motor skills to accomplish the same tasks.

Principles of Universal Design

1. EQUITABLE USE. The design is useful and marketable to persons of diverse abilities.
2. FLEXIBILITY IN USE. The design accommodates a wide range of physical preferences and abilities.
3. SIMPLE AND INTUITIVE USE. Use of the design is easy to understand, regardless of user's experience, knowledge, language skills, or current concentration level.
4. PERCEPTIBLE INFORMATION. The design communicates necessary information effectively to the user, regardless of ambient conditions or the user's sensory abilities.
5. TOLERANCE FOR ERROR. The design minimizes hazards and the consequences of accidental or unintended actions.
6. LOW PHYSICAL EFFORT. The design can be used efficiently and comfortably and with a minimum of fatigue.
7. SIZE AND SPACE FOR APPROACH AND USE. Appropriate size and space is provided for approach, reach, manipulation and use, regardless of user's body size, posture, or mobility.

Figure 2. Principles of Universal Design (8).

In place of the small syringes and Pasteur pipets commonly used to dispense 1- and 2-mL volumes, we have provided a variety of syringes fitted with disposable pipet tips. A plastic bottle cap glued to the top of the plunger (Figure 3), enlarges the grip. Once the student is introduced to the system and provided with an assortment of syringes, caps, pipet tips and a tube of suitable glue, he assembles his own tools, as needed.

In the General Chemistry curriculum, students are taught to use a volumetric pipet with a 60-mL pipet bulb. For a student with a hand tremor, a screw-type pipettor was easier to use. To further damp the effect of the tremor, foam pipe insulation (Figure 4) was fitted to the pipettor and some of his other tools.

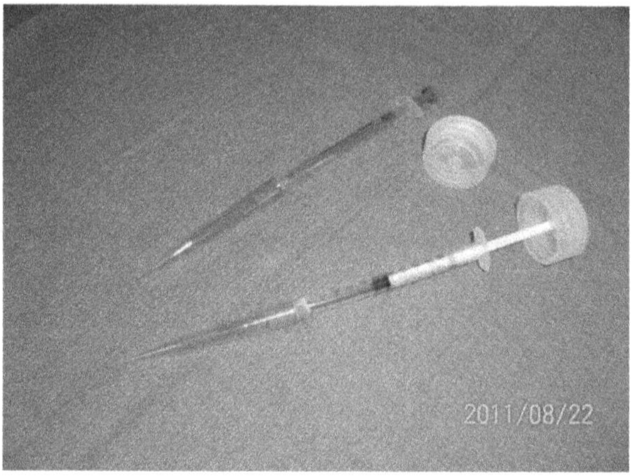

Figure 3. Syringes adapted with larger handles. Reproduced with permission from ref (6). Copyright 2013, Elsevier.

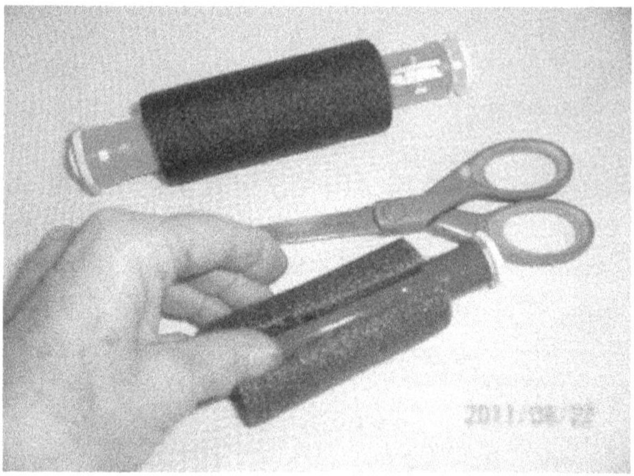

Figure 4. Pipe insulation provides cushioned hand grips. Reproduced with permission from ref (6). Copyright 2013, Elsevier.

Another hand grip modification involved enlarging handles on the stop-cocks for burets and separatory funnels (Figure 5). Embedding the small handle into a large knob (with two-part epoxy cement) gave the student more control of these fine motor tasks. A white line was added to indicate the handle position.

Costs of these projects were minimal and covered by CHEM. Some of these adapted tools have now been used by several students. Most materials were found at a local hardware outlet.

Figure 5. Extended handles for buret and separatory funnel stopcocks.

Industrial Sit–Stand Stool

An industrial sit–stand stool allows a worker to avoid standing all day, while remaining close to the work (Figure 6). The triangular base design is stable and gives a worker a chance to place both feet close to a cabinet where no knee well is provided. The seat swivels 16° in either direction and returns to center automatically. The seat tilts forward up to 15° and the seat height adjusts from 23½" to 33½". These stools have now been used in our labs by several students who found standing for the three-hour laboratory period difficult. One of these had chronic back pain; another was seven months pregnant.

Figure 6. Industrial sit–stand chair. Reproduced with permission from Lyons, LLC.

Exposed Toes

Several times we have seen students with broken foot bones that were nearly healed. In each case, the student was ready to stand and walk without crutches but was confined to an orthopedic boot. With the student ready to attend laboratory course in all other respects, we concerned ourselves with protecting the toes that were left exposed to chemical splash and to sharps, such as broken glass. A normal shoe could not be worn, as there was only room inside the boot for the thickness of a sock.

The remedy in one case was a $12 pair of ballet slippers. These very thin slippers are made of soft leather. They come in a wide range of sizes and in the classic ballet colors of black, white and pink. Figure 7 shows the ballet slipper in the orthopedic boot plus a view of the thin slipper on the other foot. For lab work, the student wore a leather boot on the uninjured foot. The heel of the boot gave similar support and an even hip height.

Figure 7. Ballet slipper protects vulnerable toes in a walking boot. Reproduced with permission from ref (6). Copyright 2013, Elsevier.

It's important to notice that these slippers are not the specilaized shoes that allow a ballerina to dance *en pointe.* Neither are they "ballet flats," which are much sturdier shoes for general use. Available from dance supply stores, ballet slippers are constructed mostly of a single thin layer of leather; the sole is one or two thin layers. Being soft and flexible, the slipper conforms to the foot and has no heel. The whole slipper fits inside the boot and adds very little thickness while protecting the student's toes from spills and cuts.

We have also experimented with a number of other toe covers. One of the more successful was a plastic beaker cut in half (Figure 8). Sharp edges were trimmed and covered with duct tape. Further consideration showed that plastic beer cups were readily available and easier to trim. Plastic cups should be tested against the solvents students will use (and may spill). With a supply of appropriate cups and tape, several students have been able to reconstruct the protective cover as needed.

Figure 8. Half a plastic beaker protects exposed toes. Beer cups also work well. Sharp edges are smoothed with duct tape.

Low Vision in General Chemistry

Our most recent experience with low vision was a student with very restricted vision who benefitted from enlarging many items in her laboratory environment: large-scale glassware was provided, along with larger quantities of reactants, to produce larger, more visible quantities of chemical products. Bottle labels, exams and printed materials (such as the color chart for pH test strips and a periodic table) were enlarged. The student used racks and boxes to help her organize her drawer and save time. In a more complex chemical environment, scaling up a reaction would require consideration of the energies involved and evaluation of the increased risk.

Extra time was authorized for this student and she worked in two sections with two TAs. A directed assistant was employed and cautioned to respond only to her explicit directions. Digital models of a pH meter, thermometer and spectrophotometer were set aside for her use. High contrast versions of some items, such as graduated cylinders, were readily available. For other items, adding colored paper tape or a contrasting background was helpful in viewing and reading liquid levels.

An unusually useful tool was the student's electronic reading magnifier. It was designed for enlarging book or magazine type and had a freeze-frame function that was useful for instrument readings. Once the image was captured, it could be held on the screen, then viewed with a variety of backgrounds and contrast settings.

Low Vision in Organic Chemistry Laboratory

For a student with such restricted vision, the micro- and semi-micro-scale glassware used in our Introductory Organic Chemistry program presented a real challenge.

Working closely with the student, we used the methods we learned in General Chemistry Laboratory, plus a few new ones. We extended a loan of one of our lecture-scale model kits and provided larger glassware. As part of her on-going work with the California Department of Rehabilitation, our student acquired an

Acrobat HD projection system (Figure 9) which worked with a free-standing computer monitor. The student could focus the camera on the chalkboard and demonstrations, viewing the flat-screen monitor at her workbench. The Department of Rehabilitation supplied this instrument, which became the property of the student. We made a splash-resistant cover (a lab coat) for the camera and a loose plastic cover for the monitor.

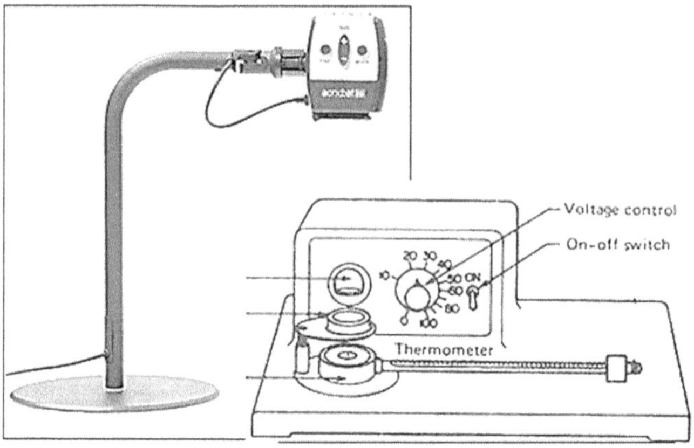

Figure 9. Fisher-Johns melting point apparatus with projection reader (monitor not shown). Reader image reproduced with permission from Enhanced Vision. Image of the melting point apparatus reproduced with permission from ref (9). Copyright 2014 John Wiley & Sons.

One of the key techniques for assessing the purity of an organic solid is determining the melting point. In our laboratories, the melt is usually observed in a few small crystals held in a capillary tube with an inside diameter of approximately 1 mm. The student places the tube into a heating block, observes it in a lighted, magnifying window and records the temperature range in which the crystals melt.

Multiple attempts to help this student see crystals melting in the capillary melting point apparatus failed. We tried a web camera, a cell phone camera, and a bore scope among other ideas. Finally, the instructor located an old stage-type Fisher-Johns melting point apparatus which was common in teaching laboratories in the 1950s through the 1970s. This model uses a larger crystal sample in a larger observation and magnification area. We replaced the thermometer with a digital model. We borrowed a page of instructions from a school that still uses the instrument, but the instrument manual still eludes us.

The old melting point apparatus, combined with the new projection system (Figure 9) worked well enough that the student observed the melt and recorded the temperature. Sometimes, reverting to "low tech" (or "old tech") is a useful area to investigate.

Wheelchair Chemistry

Over the course of three years, several students who use wheelchairs completed a General Chemistry Laboratory course. Two of them successfully completed the Introductory Organic Chemistry Laboratory course as well.

A 2001 laboratory renovation included accessible workstations and chemical hoods in several of our laboratories (Figure 10). These have 34-inch high counters in place of the standard 36-inch counters and a knee well is left open to allow access from a seated position. Some additional modifications were made to bring services within reach of a seated student (Figure 11).

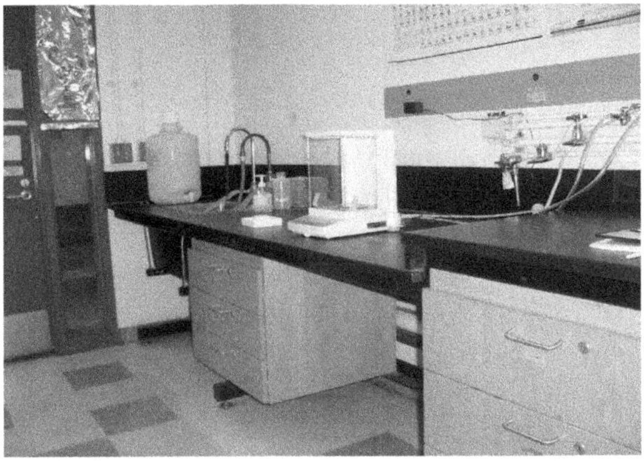

Figure 10. Accessible workstation and sink area.

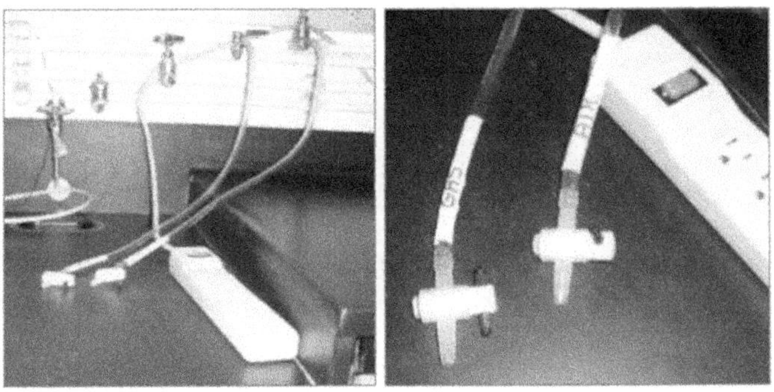

Figure 11. Temporary modifications helped a seated student reach utilities. Reproduced with permission from ref (6). Copyright 2013, Elsevier.

A too-deep sink was modified by inserting a shallow steel sink, which sits on the rim of the larger sink, and adding long handles to the water faucet (Figure 12). Some additional modifications were as simple as supplying mirrors to a student who couldn't see an instrument from her chair (Figure 13). The modified workstation has now been used by students ranging from frail and petite to tall and athletic. All of these students completed the courses.

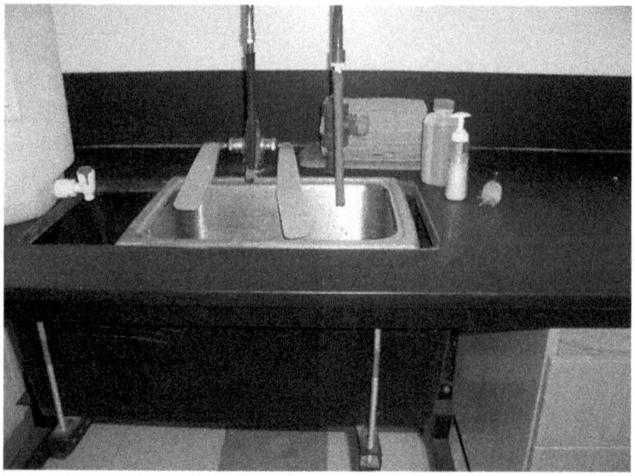

Figure 12. A too-deep laboratory sink modified by adding a shallow sink and long handles. Reproduced with permission from ref (6). Copyright 2013, Elsevier.

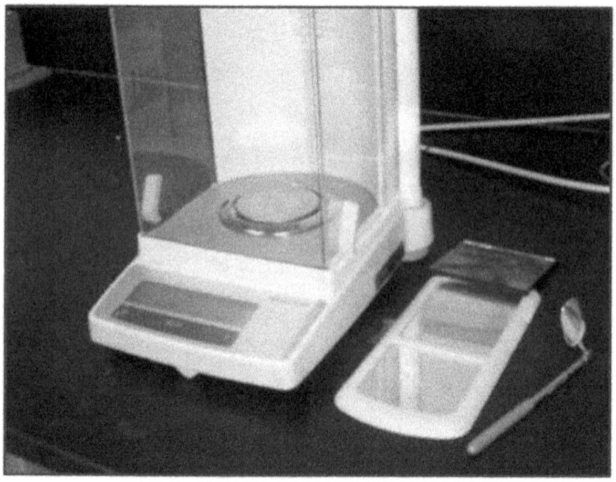

Figure 13. Mirrors help a seated student see an instrument. Reproduced with permission from ref (6). Copyright 2013, Elsevier.

All of these modifications were done in the CHEM mechanical shops and most were done with parts on hand or at small expense.

Summary

In all our work to accommodate students with disabilities in laboratory courses, our most useful tools have been the student's previous experience, our documented procedures, the creativity and diversity of our professional team, our partnerships with professionals across campus, and a general willingness to identify and think beyond our assumptions.

In each case, we were most successful when we adhered to our guiding principles, made no assumptions ahead of time, and worked openly with everyone concerned.

References

1. Americans with Disabilities Act of 1990 as Amended, Pub. L. No. 101-336, 104 Stat. 328; 1990. https://www.ada.gov (accessed May 7, 2017).
2. Rehabilitation Act of 1973. www.dol.gov/oasam/regs/statutes/sec504.htm (accessed May 7, 2017).
3. National Alliance on Mental Illness. Mental Health by the Numbers. https://www.nami.org/Learn-More/Mental-Health-By-the-Numbers (accessed October 16, 2017).
4. Kennedy, S. M.; Boval, J. Accommodations for Laboratory Students with Low Vision. Presented at 252nd National Meeting of the American Chemical Society, Philadelphia, PA, August 2016; Paper CHAS 11.
5. Kennedy, S. M. Choosing the Right Tool: Accommodations for Students with Low Vision in Chemistry Teaching Labs. Presented at the 248th National Meeting of the American Chemical Society, San Francisco, CA, August 2014; Paper CHAS 19384.
6. Kennedy, S.; Palmer, J.; Boval, J. Adaptive tools for laboratory students with disabilities: Case studies from UC San Diego. *J. Chem. Health Saf.* **2013**, *20*, 35–42.
7. Howard, L. , Potts, E. A. *Including Students with Disabilities in Advances Science Classes*; NSTA Press: Arlington, VA, 2013.
8. The Center for Universal Design. *The Principles of Universal Design*, Version 2.0; North Carolina State University: Raleigh, NC, 1997.
9. Zubrick, J. W. *The Organic Chem Lab Survival Manual: A Student's Guide to Techniques*, 9th ed.; John Wiley & Sons, Inc.: Hoboken, NJ, 2014.

Chapter 9

Finding Accommodations That Work

Carl E. Heltzel*

ACS National Committee Member, Chemists with Disabilities, 102 Bobolink Drive, Seneca, South Carolina 29672, United States
***E-mail: carlheltzel@gmail.com**

The previous chapters provided information regarding students or employees with disabilities and examples of accommodations for helping them obtain the same instruction in laboratory concepts and skills as non-disabled persons. This final chapter aims to clarify some specific issues and present a firsthand account of how one might go about finding accommodations that are successful, beneficial to the entire class/industrial lab, how to get started in this process, and suggestions for assessing the success of an accommodation. Section headings include Overview, Why This Is Important, Where to Begin, Federal Agencies, World views on Accessibility, Creating a Reasonable Accommodation, Self Disclosure, Virtual and Remote Labs, Assessing Success of Accommodation, and Conclusion. The References list at the end of the chapter offers an enormous amount of information and resources.

Overview

This book has provided information on many types of disabilities, laws and regulations that must be followed, and case studies describing various laboratory accommodations for persons with disabilities. Despite the fact that students with disabilities (SWD) are achieving successful high school education and entering post-secondary institutions at higher rates than ever, SWD are graduating at a rate lower than that of their non-disabled peers; for some time, it has been suggested that the primary reasons for this disparity include: 1) preparation of students with disabilities; 2) access to facilities, programs, and equipment; and 3) acceptance by educators, employers and co-workers (*1*). In reality, the reasons for the high attrition rate for SWD are many and an equal number of strategies to remediate the issue have been proposed; Universal Design and shifting the focus from providing accommodations for individual SWD to completely removing barriers in the environment in order to provide universal access, Critical Disability Theory and its alternative viewpoint for reviewing accessibility and encouraging full participation of campus stakeholders for resolving accessibility issues, digitization and electronic library resources and removing barriers for persons with visual impairments, ... and the list goes on and on. The higher education climate going forward is a complex and tangled web of dwindling resources, federal legislation, need for accountability, recruiting, and retaining students and is far too complex to give a comprehensive review in this book.

But for the approximately 700,000 students with disabilities enrolled every year in 2- and 4-year colleges (*2*), as well as nearly 18% of the U.S. workforce with a disability (*3*), and anyone else who wants to study chemistry, we want clear and concise information to be readily available. In this final chapter, it is appropriate to reiterate the fact that all aspects of making accommodations should not fall upon the instructor or employer alone. This is not their job and they cannot be expected to know what the best course may be for making an accommodation in the lab. In addition, the hazards associated with working in a lab are very different than in an office or a classroom. Successful accommodations, also referred to as "productivity enhancers" are made when all of the stakeholders are involved in the decision process, to decide whether or not an accommodation is warranted, and precisely what the accommodation should be.

The laboratory is a very specialized setting and qualified SWDs, unfamiliar with the accommodation process and their associated rights, might easily be swayed into another direction. This sounds like an unfortunate and potentially illegal situation, however, in reality it is a relatively common occurrence. But, armed with the proper information-*forewarned is forearmed*-all qualified individuals should have the same access to laboratory instruction as anyone else, and may very well excel in a science career.

Armed with the proper information–forewarned is forearmed– all qualified individuals should have the same access to laboratory instruction as anyone else, and may very well excel in a science career.

Also, the final chapter is an appropriate place to review and consider the information from the preceding chapters in terms of: 1) How does one find accommodations that are effective, and 2) How can the effectiveness of the accommodations be evaluated? This book is filled with information, both technical and anecdotal, and citations for finding the original sources for the information. In this final chapter, we would like to describe some first-hand observations of laboratory accommodations for students with disabilities that have been successful, as well as some that have not.

The references section in this chapter will provide you with web sites and additional resources that you might find helpful in your endeavor to create an atmosphere of inclusiveness in your laboratory.

Why This Is Important

Today's global economy and surge of technology is creating a demand for qualified persons with STEM skills, and persons with disabilities are significantly underrepresented in the sciences-both in academia and industry. In fact, persons with a disability are underrepresented in the workforce in general. In 2016, the U.S. unemployment rate for persons with a disability was 10.5%, about twice that of those with no disability (4.6%), according to the Bureau of Labor Statistics (BLS) (*3*). Furthermore, in 2016, across all levels of education, persons with a disability were much less likely to be employed than their counterparts with no disability (*4*).

Also consider the increasing demand for healthcare professionals, all of which will require some chemistry to attain their proper degree or certification. Employment of healthcare occupations is projected to grow 18 percent from 2016 to 2026, much faster than the average for all occupations, adding about 2.3 million new jobs (*5*). Targeting and supporting students with disabilities by encouraging them to seek a STEM education trajectory may help to fill the need for a technologically skilled workforce because they may bring unique and innovative problem-solving skills to industry and academia. ***After all, persons with disabilities have often had to devise numerous innovative ways to simply navigate life.***

There are also faculty and staff members who are confused about all aspects of making accommodations, an issue with which we sincerely hope to offer help and advice in this book. We want this information readily available not only because supporting persons with disabilities is the *right* thing to do, it is the *only* thing to do; it is required by law.

Where To Begin

When finding accommodations that work, there is no need to reinvent the wheel. That is, the first step in this direction might simply involve asking colleagues across your institution what efforts they have made in terms of making their laboratories more accessible. These inquiries shouldn't be restricted to chemistry or chemical engineering labs; lab settings are spread all across campus. Consider the Art Department, Biology, Physics, Automotive Technology, Welding, etc.-they all have labs and associated risks.

A visit with your Office for Students with Disabilities (OSD, or whatever term is used at your organization), might uncover valuable best practices as well as instances where attempts to make accommodations were unsuccessful. Being familiar with the OSD staff can be very helpful when a person with a disability starts to work in your lab. If your institution has benchmarked organizations, consider approaching them for information. I found that instructors, OSD staff, lab coordinators, and administrators at research universities, colleges, and two-year institutions are more than willing to share their experiences with successful implementations, adoptions they have planned but not yet executed, as well as those that were unsuccessful.

Furthermore, I spoke with people from various institutions that have not considered an accommodation process because they have not yet been confronted with a single SWD; we would like to suggest that it is better to be proactive in this regard, rather than waiting until the need arises. An institutional structure for conducting the assessment of an accommodation should already be in place as well.

Check What Federal Agency Labs Are Doing

Federal agencies have a legal obligation to provide reasonable accommodations for their employees and job applicants with disabilities, including agencies where laboratory-based employees are hired. When making a determination on whether a reasonable accommodation can be provided, the decision-maker should not consider just the budget in the office where the employee works but consider the budget of the entire agency. For this reason, denial of a reasonable accommodation request based on a financial undue hardship determination should be extremely rare (*6*).

The National Aerospace and Science Agency (NASA) (*7*) has added verbiage to their newest request for proposals; a sentence stating that "NASA recognizes and supports the benefits of having diverse and inclusive" communities and "fully expects that such values will be reflected in the composition of all proposal teams." NASA has taken a tangible step towards enhancing diversity by providing funds to bring outside scientists to many of their in-progress missions, allowing the agency to add expertise to their research teams as well as improve inclusiveness.

Other governmental agencies are taking steps to make their laboratories accessible to their employees, and a simple web search will turn up programs similar to NASA's. Our federal agencies have expended significant effort into making laboratories accessible, so it makes sense to see if their results might be useful in your unique situation.

With all of the resources available, there should be no one in a position of laboratory oversight that hesitates to talk about accommodations for SWD–the legal responsibilities of the ADA, the real or perceived burden placed on an instructor when a SWD enrolls in a course, what steps *can* be taken, and when specific steps *must* be taken.

Accessibility Is Not Just an American Issue

Accommodations and guaranteed access for persons with disabilities are not topics restricted to labs in the U.S. For example, in Ontario, Canada, frameworks are outlined in the **Ontario Human Rights Code** (the Code) and the **Accessibility for Ontarians with Disabilities Act 2005** (AODA), both of which aim for sweeping systemic change (*8*). The Code is a "rights-based framework with stronger, but more general procedural and substantive obligations" than the AODA. The Code expects immediate compliance, and provides an individual complaint mechanism to help enforce obligations. Further information on the rights of persons with disabilities, including aspects regarding education, requirements of accessibility, and deadlines for meeting these requirements can be found at: www.ontario.ca/page/accessibility.

In August 2017, I met with Sir John Holman, founder of the United Kingdom (UK) National Science Learning Center and President of the Royal Society of Chemistry (RSC), to discuss accommodations for SWD in the UK. Sir Holman reported that, in England, there are clear lines of responsibility for ensuring that students with disabilities receive the help they require, and the topic is gaining attention. Individuals have been protected from discrimination by the **UK Disability Discrimination Act** (*8*), enacted in 1995. This legislature imposes similar requirements to those requirements under the ADA. However, the law does not require retrofitting; so a number of labs built prior to this legislative effort remain difficult to navigate for scientists with mobility issues.

Other issues that still need attention include the transition of SWDs to the workplace. The UK **STEMM Disability Advisory Committee** (*9*), formed in 2011, along with the RSC and other scientific groups across the UK, made a push to improve funding to scientists with disabilities. Another outcome of this group's efforts is the **Royal Holloway STEM Disability Transition Group** (*10*), which endeavors to facilitate greater numbers of young people with disabilities to develop STEM skills and find work in technological fields. Towards that end, the group created the **Employers Toolkit**, which provides practical information, sources of support, and case studies to help employers develop more inclusive and accessible apprenticeships.

In Japan, the **Act for Eliminating Discrimination Against Persons with Disabilities**, enacted in 2013 and came into effect in 2016, prohibits unfair discriminatory treatment by governments and private entities based on disability. The **Amended Act on Promotion for Disabled Person's Employment** further requires that Japanese government-related agencies provide reasonable accommodation to enable the elimination of social barriers, unless the expense is "excessive." This legislation requires that public universities provide reasonable accommodations to students with disabilities (*11*). On a global scale, 2016 marked the 10th anniversary of the adoption of the United Nation's **Convention on the Rights of People with Disabilities and Optional Protocol** (*12*), to uphold the dignity and protect and promote the rights of persons with disabilities. The treaty was adopted in 2006 and was opened for signature in 2007, and came into force in 2008. This is the first comprehensive human rights treaty of the 21st century, and had the highest number of signatories in history to a UN Convention on its opening day. The ***Convention clearly reaffirms that the rights of persons with disabilities are human rights*** and serves to strengthen respect for these rights. As of October 2016, 87% of the UN Member States have ratified the Convention and 47% have ratified the Optional Protocol. The United States has not ratified either of the documents. Translations of the Convention into national sign languages can be found on the World Federation of the Deaf web site at: https://wfdeaf.org/human-rights/crpd/.

The United Nations declared that, from 1983 to 1993 was the **International Decade of Disabled Persons**, and that every year on December 3rd is the **International Day of Persons with Disabilities** (*13*). Each year, the day focuses on a different issue or theme. Simply searching the web yields significant information and discussion points for creating an atmosphere of inclusiveness in the lab for both students and professional chemists.

Creating a Reasonable Accommodation Process

Creating an accommodation is a very individualized process because of the many different needs and the wide variety of venues. High schools, two-year colleges, universities, and industries can have vastly different laboratory environments. Consider a graduate student with a disability; like all graduate students, success is measured by productivity and output (*14*). Any length of time taken to devise an accommodation is considered time "lost." This sort of attitude would not be acceptable if we were discussing safety training that all graduate students need. Time must be a consideration when creating an accommodation. Ideally, every institution that has a laboratory on site should take steps to craft an accommodation process before the need arises. There is strong evidence that an inclusive laboratory, in addition to the classroom or other training facilities, benefits not only SWDs, but everyone involved (*15*). With the passage of the **ADA Amendments Act** (ADAAA) of 2008, creating a reasonable accommodation process must be consistent, fair, and equitable, which is the legal minimum standard.

The faculty member, lab instructor, or employer responsible for making science laboratories accessible is not expected to accomplish this alone. The institution, OSD staff, and student/employee with a disability all have roles in identifying needs and participating in discourse regarding making the lab accessible. It is important that university level SWD develop self-advocacy skills, become more aware of their individual needs in diverse environments, and learn how to communicate these needs to faculty and disability services staff (*16*). When faculty demonstrate an understanding of the accommodation process and use inclusive teaching methods, SWD feel more likely to succeed.

Self-Disclosure

There is a requirement that certain contractors working with the federal government actively ensure that 7% of their workforce is made up of people with disabilities (*17*). This creates a situation where such employers need to ask applicants and employees to voluntarily self-identify as an individual with a disability. Requesting this information will help contractors reach their mandated 7% goal, but in the workforce as well as in academia, many job applicants/students do not disclose their disability. In order to achieve inclusiveness in the workplace we will have to understand the reasons why people don't self-disclose. While fear is thought to be the primary reason, using focus groups with employees with disabilities, A Springboard, LLC report suggested that the main reason is that most people simply don't know how (*18*). Students and employees are unsure what and how much information to share, and with whom should they discuss this. SWD want assurances regarding an anticipated outcome–not a guarantee of a good grade or job security, but that they will be accepted as equals, go through the same steps, learn the same concepts and skills, and be as prepared as non-disabled students as they navigate their careers. Universities and companies would be well served to provide guidance to clarify the responsibilities of everyone involved. This can be accomplished by adding a disability disclosure information page to their website, application process, and class syllabi.

Providing a concise, easy to understand disability disclosure guide on a college course syllabus or a company's website helps students, job candidates, and employees feel more informed, better equipped, more confident-and more likely-to disclose their disability. The document should be created in collaboration with the OSD staff, keeping personal information secure.

Regarding the posting of a disability disclosure on the web, ideally, all websites should adhere to the **Web Content Accessibility Guidelines** (WCAG) global standards. **WCAG 2.0** and the recently published **WCAG 2.1** were developed through the **World Wide Web Consortia** (W_3C) (*19*) with the help of organizations and individuals from many countries and covers a wide range of recommendations with the primary goal of providing a single standard for web content accessibility that meets the needs of individuals, organizations, and governments around the world.

There is so much information available that it may seem overwhelming. As the responsibility of accessibility does not fall upon an individual faculty member or a corporate manager, WCAG documents help explain how to make web content more accessible to people with disabilities. Web "content" generally refers to the information in a web page or web application, including:

Information and aspects for making accessible web pages include:

- Page title
- Image Text Alternatives ("alt text"; pictures, illustrations, charts, etc.
- Text: headings, contrast ratio, resize text
- Interaction: keyboard access, visual focus

Also more general information and a basic checklist

Following the WCAG guidelines will make content accessible to a wider range of people with disabilities, including blindness and low vision, deafness and hearing loss, learning disabilities, cognitive limitations, limited movement, speech disabilities, photosensitivity and combinations of these. Like many of the suggestions in this book, as well as those in the ACS book, *Teaching Chemistry to Students with Disabilities*, following these guidelines will make your web content easier to understand for all users (*20*). For more information, including where to start, web accessibility principles and guidelines, and more technical details, there is a wealth of information available to all at www.w3.org.

An accommodations committee, good communication between parties, a well-thought-out process that includes how requests are made, who can make them, criteria for providing the accommodation, turnaround time, process for submitting a claim of unfairness-all of these aspects make up a viable accommodation process. When expectations, deadlines, rules, and regulations are clearly stated, it is less likely that a problem will arise.

Integration of the Accommodation with the Course Instruction

Keep in mind that accommodations are for *qualified* individuals; any student requesting an accommodation must have passed any prerequisite courses and have the intellectual capacity to succeed in the class. When instructors, lab managers, chemical hygiene professionals, and lab assistants are all instructed in best practices regarding supervising persons with disabilities and making accommodations, and everyone knows what part(s) they are responsible for, the process should go well. Ongoing training in this area, coordinated with the OSD staff, is central for integrating the student's accommodation into the course with as little disruption as possible. The student or employee with a disability wants to feel included and rarely uses a disability to gain an advantage over other people in the lab. Some considerations for helping to integrate the accommodation into the 'regular' class or 'normal' day-to-day routine in a professional lab might include:

Protective Clothing

Open Style Lab (OSL) is a nonprofit organization dedicated to making style and clothing accessible to people of all abilities. One of their success stories is the Versacoat, an adaptive, multi-season jacket that provides full weather protection and durability during propelled movement or wheelchair transfer. It has many pockets that can securely hold a user's everyday necessities. The lab coat also has a back hemline that ends at the wheelchair seat, and an elongated front for leg coverage, which is ideal for wheelchair users (*20*).

Color Vision Deficiency Aid

Worldwide, 1 in 12 men and 1 in 200 women are affected by some form of color vision deficiency (CVD), the current term for "color blindness." It is also known as Daltonism, as John Dalton published the first scientific paper on color blindness. People with this condition see colors-they are not blind to them-but their ability to see them is reduced. The most common is red-green color blindness; the red and green photopigments overlap more than normal in the eyes of people with this condition. This not only causes confusion with red and green colors, the entire color spectrum can be confusing. CVD can limit a person's ability to observe color changes in certain chemical reactions such as an indicator-based acid-base titration.

Companies now offer glasses that filter out specific colors to alleviate this disability. Enchroma has a website where one can take an eye exam for CVD and learn about new technology used to reduce the impact of red-green color blindness by using special lenses that target specific photopigments (*21*). Their lenses, made using a technique called 'multinotch' filtering, cuts out sharp wavelengths of light to enhance specific colors, allowing people with CVD to distinguish red and green colors. It is important to note that the glasses are not effective for all persons with red-green color blindness and that they do not provide 100% accurate color vision. Still, this is an accommodation that would be beneficial to anyone with CVD that works in a lab where the ability to discern colors is important.

Concerned that normal plastic goggles worn over these special lenses might interfere with their operation, thus making them useless for lab accommodations, I made some inquiries. The CEO of Enchroma stated, "Its [wearing goggles over their glasses] not ideal due to the weight and some internal reflections between the two sets of lenses, but it doesn't interfere." He added, "Alternatively, we would need to make a custom pair of glasses in an ANSI Z87.1 frame (safety frame)" (*21*). I went a step further and asked my nephew, who has red-green color blindness, to test the effectiveness of the lenses while wearing typical splash goggles. Figure 1 shows him using a universal pH indicator strip to see if a solution was acidic (it turned red) or not.

Figure 1. A young student with red-green color blindness using special glasses as he inspects the color of a universal pH indicator test strip. Without the glasses, it would be impossible for him to discern red and green colors. Photo courtesy of Deborah Bone

Virtual Laboratory Experiments

An online search for "virtual lab experiments" turned up an ACS page dedicated to disseminating resources for virtual lab exercises: "Chemistry moves from the lab and the classroom to the computer, as working in a virtual chemistry laboratory and viewing simulations provide additional ways of learning chemistry (*22*)." Every issue of *Chemistry Solutions*, the periodical of the American Association of Chemistry Teachers (AACT), contains a lab simulation (*23*). The ACS.org page inventories a selection of simulations from past issues. I tried the "Half-life Investigation" activity and it was easy to visualize and understand the concept of a half-life by watching a virtual, unstable radioactive material decay over time until the sample became stable. The data was plotted for me automatically. This is a great example of using virtual labs to learn a concept, especially for visual learners; it is available to anyone, of any ability, and at any time–and of course this experiment could never be safely done in a college lab.

ChemCollective, a collection of virtual labs as well as scenario-based learning activities, was created by a group of faculty and staff at Carnegie Mellon University who are interested in using, assessing, and creating engaging online activities for chemistry education (*24*). Again, I found free access to many sophisticated virtual lab experiments.

LabsLand, a relatively new organization based in Spain, offers various services where subscribers can access real labs in real time. As of August 2017, their lab experiments were limited to physics and optics, but they are moving towards crafting remote access, real time–in real labs–chemistry experiments (*25*). The physics lab experiments can allow a person, via a computer, to see a real lab, and in real time, the student can activate various instruments and generate real data. This is not a virtual lab or a simulation, it is a true experiment where the student remotely operates real equipment–but the lab may actually be in Spain or Brazil. As rapidly as technology is advancing robotics, it seems realistic that within a short period of time, students will have automated pipets draw up a solution and move it to another flask and release the aliquot–simply by pushing a few buttons. Will it matter if the lab instrument is three feet or three thousand miles away?

A point for discussion is now upon us concerning criteria that will define what virtual or remote lab experiences qualify for college credit so that people of all abilities and socioeconomic backgrounds may obtain degrees in chemistry. It is an exciting time to be a chemist!

Assessing the Success of the Accommodations

With every strategy, approach, adaptation, etc., it is important to verify the success of steps that were taken. This is true for all programs; if there is no attempt to assess whether or not a program is having an intended effect, that everyone being affected by the program is benefiting, and that it is not cost prohibitive, then that program stands a high chance of failure. Formative as well as summative assessments would be ideal, as both ‘disability’ and ‘accommodation’ can be moving targets, so periodic examinations about the accommodation will determine if some adjustment is needed. Specific questions one might ask when creating a summative matrix for assessing the success of a laboratory accommodation (“employee” can replace “student” as needed):

- Were all stakeholders included in the decision-making process and was the accommodation added without disrupting the regular class?
- After the accommodation was set up, did the student use it regularly?
- Did all the students have the same experience/learn the techniques required?
- Was there a change in the SWD’s attitude concerning chemistry and their learning and performance?

- Did the accommodation allow the SWD to feel included as every other class member?
- Was the SWD able to attain the same level of skills and learning per the syllabus as the rest of the class?

Assessment will take effort requiring the collection and analysis of data. This should not be the instructor's responsibility. The collaborative team should decide how, when, and who will collect the data. If the accommodation does not appear to be effective, try something else. Collect only data that is needed to steer action towards realistic, achievable goals. Keep lines of communication open and make sure all stakeholders understand their roles.

With all of the considerations mentioned in this section, along with freely available resources, creating accommodations and assessing their success can be a straightforward experience that benefits not only the student with a disability, but everyone in every lab setting.

Conclusion

Collaboration is the key to success when finding accommodations that work. An instructor must be willing to even the playing field for students of all abilities. Technology is rapidly advancing at a pace where it might behoove faculty to keep an open mind about accommodations, be they modified faucet handles or remote access lab experiments. Lab instructors, teaching assistants, and corporate lab managers must all be trained in best practices, in collaboration with their OSD staff. The SWD should exhibit a motivated effort to use the accommodation, which, in turn, has the added bonus of bolstering his/her feeling of self-sufficiency. Designing a successful accommodation involves full participation of all stakeholders including the administration, and in some cases, the student's parents. The institutions that do have a program for making accommodations for SWD and a matrix for assessing success sends out a clear signal of a mature organization that looks for continuous improvement. In today's climate of funding shortages and declining enrollments, when an adoption may mean cost to a department, a valid assessment is a good tool for showing accountability.

Traditionally, there has not been much collaboration between instructors in the STEM fields of postsecondary education, commercial laboratory management, and disability professionals. More attention should be focused on the technological advances that are taking place whether we agree with their use or not. More funding should be available for determining the steps needed for enhancing inclusion and recruiting students with disabilities into STEM fields of study as well as determining and disseminating the fundamental best practices. The availability of funding for gaining insight into all of these areas might provide an incentive for disability professionals and postsecondary education instructors to work together. It's in the best interest of the students (disabled or not), instructors, administrators, and OSD staff to work together as the future descends upon us and the paradigm that laboratory instruction is based upon shifts once again.

References

1. Burgstahler, S. *Increasing the Representation of People with Disabilities in Science, Engineering, and Mathematics*; University of Washington, Disabilities, Opportunities. Internetworking, and Technology: 2017. http://www.washington.edu/doit/increasing-representation-people-disabilities-science-engineering-and-mathematics (accessed November 19, 2017).
2. Raue, K.; Lewis, L.; Coopersmith, J. *Students With Disabilities at Degree-Granting Postsecondary Institutions*; NCES 2011-018; U.S. Department of Education, National Center for Education Statistics: 2011. https://nces.ed.gov/pubs2011/2011018.pdf (accessed November 19, 2017).
3. U.S. Department of Labor, Bureau of Labor Statistics, Economic News Release, Persons with a Disability: Labor Force Characteristic Summary; USDL-17-0057; June 2017. www.bls.gov/news.release/disabl.nr0.htm (accessed November 9, 2017).
4. U.S. Department of Labor, Bureau of Labor Statistics, Economic News Release, Persons with a Disability: Labor Force Characteristic Summary; USDL-17-0057; June 2017. https://www.bls.gov/news.release/disabl.t01.htm (accessed November 9, 2017).
5. U.S. Department of Labor, Bureau of Labor Statistics, Occupational Lookout Handbook; 2017. https://www.bls.gov/ooh/healthcare/home.htm (accessed November 9, 2017).
6. U.S. Equal Employment Opportunity Commission (EEOC). Enforcement Guidance: Reasonable Accommodation and Undue Hardship Under the Americans with Disabilities Act, Notice Number 915.002; 2002. https://www.eeoc.gov/policy/docs/accommodation.html (accessed November 9, 2017).
7. *Guidebook for Proposers Responding to a NASA Funding Announcement*; NASA: 2017. https://www.hq.nasa.gov/office/procurement/nraguidebook/proposer2017.pdf (accessed November 9, 2017).
8. UK Disability Protection Legislature. www.gov.uk/rights-disabled-person (accessed September 25, 2017).
9. Future Directions in STEMM for People with Disabilities; Royal Society of Chemistry: 2016. http://www.rsc.org/events/detail/21249/future-directions-in-stemm-for-people-with-disabilities (accessed November 8, 2017).
10. STEM Disability Transition Group Conference 2014: What's So Special About STEM? Conference Report; Royal Holloway, University of London. https://www.royalholloway.ac.uk/studyhere/documents/pdf/educational advisors/whatssospecialaboutstemconferencereport2014.pdf (accessed November 8, 2017).
11. Matamura, A. New Disability-Related Law of Japan: Law for Employment Promotion of Persons with Disabilities; 2013. www.jldd.jp/gtid/news_form/

employment-promotion-law-of-PWD20130619.pdf (accessed September 25, 2017).
12. United Nations General Assembly Session 61, Resolution 106. Convention on the Rights of Persons with Disabilities, A/RES/61/106; 2006. http://www.un-documents.net/a61r106.htm (accessed November 9, 2017).
13. *The United Nations and Disabled Persons, The First Fifty Years*; Chapter VII: United Nations Department of Economic and Social Affairs, Division for Social Policy and Development Disability; United Nations Enable: 2003.http://www.un.org/esa/socdev/enable/dis50y00.htm (accessed November 8, 2017).
14. Sukhai, M. A.; Mohler, C. E. *Creating a Culture of Accessibility in the Sciences*; Elsevier, Academic Press: Atlanta, GA, 2017.
15. *Disability and Equity in Higher Education Accessibility*; Alphin, H., Jr., Lavine, J., Chan, R., Eds.; I.G.I. Global: 2017.
16. University of Washington, Disabilities, Opportunities, Internetworking, and Technology. www.washington.edu/doit (accessed November 8, 2017).
17. United States Department of Labor, Office of Federal Contract Compliance Programs, Final Rule: Section 503 of the Rehabilitation Act. https://www.dol.gov/ofccp/regs/compliance/section503.htm (accessed November 9, 2017).
18. Vogel, N. The "R" Words: The Rules for Successfully Recruiting and Retaining Individuals with Disabilities. *Profiles in Diversity Journal*, January 12, 2015.
19. World Wide Web Consortia. www.w3.org (accessed September 20, 2017).
20. Open Style Lab. www.openstylelab.com (accessed September 20, 2017).
21. Schmeder, A. CEO of Enchroma, Berkeley, CA. Personal communication, 2017.
22. American Chemical Society Online Experiments and Tutorials. www.acs.org (accessed September 11, 2017).
23. American Association of Chemistry Teachers. www.teachchemistry.org/periodical/simulations (accessed September 15, 2017).
24. ChemCollective Online Resources for Teaching and Learning Chemistry. www.chemcollective.org (accessed September 9, 2017).
25. Orduña, P. LabsLands, San Francisco, CA. Personal Communication, 2017.

Editors' Biographies

Ellen Sweet

Ellen Sweet, MS, is the Laboratory Ventilation Specialist in the department of Environmental Health and Safety (EHS) at Cornell University. She has a Master's degree from the Rochester Institute of Technology in EHS Management and is a Certified Chemical Hygiene Officer. Ellen is responsible for the Laboratory Ventilation Management Program (LVMP) which is a comprehensive management system that balances health and safety with energy conservation strategies for research and teaching labs. She is also a member of the American Chemical Society's Division of Chemical Health and Safety (CHAS), a CHAS Executive Committee member, the chair of the CHAS Regulatory and Public Affairs committee, and a CHAS Fellow.

Wendy Strobel Gower

Wendy Strobel Gower is a project director at the Yang-Tan Institute on Employment and Disability at Cornell University. She leads the Northeast ADA Center, the Diversity Partners Project, and fee for service work to educate managers and supervisors about including people with disabilities. Wendy holds a Master's degree in Rehabilitation Counseling from the Medical College of Virginia at Virginia Commonwealth University (VCU). She has worked extensively in the application and training of issues around employment and reasonable accommodation in the workplace for people with disabilities. She has gained valuable experience in project management and project direction over the past ten years. Other areas of interest include ensuring access to work, school, and community for people with disabilities, disability legislation and its impact on services, and the identification and accommodation of the functional limitations of disabilities across the lifespan.

Carl E. Heltzel

Carl E. Heltzel is a freelance grant writer, author, and editor and a member of the American Chemical Society (ACS) Committee on Chemists with Disabilities. He has a PhD in organic chemistry from Virginia Tech, and was a post doc at the University of Hawaii. He was the Monroe Moosnick Professor of Chemistry and a Bingham Fellow at Transylvania University. He also taught at Clemson University, the University of Hawaii, Manoa, the University of Kentucky, and Chaminade University. He is a consultant to Environmental Research, Inc., and has worked as a consultant for the Ohio EPA, the California EPA and Solid Waste

Removal Board, various environmental law firms, the ACS, the Department of Homeland Security, McGraw-Hill, and the International Academy of Oral Medicine & Toxicology. He served the ACS as co-author of two editions of *Chemistry in Context*, and as a past editor of *ChemMatters* magazine. In the Lexington, KY section of the ACS, he held all of the executive positions, including Councilor. He is currently a Regional Representative for the American Association of Chemistry Teachers.

Indexes

Author Index

Subject Index

L

P

www.ingramcontent.com/pod-product-compliance
Ingram Content Group UK Ltd.
Pitfield, Milton Keynes, MK11 3LW, UK
UKHW021315130926
15127UKWH00002B/12

9 780841 232761